SpringerBriefs in Applied Sciences and Technology

Computational Intelligence

Series Editor

Janusz Kacprzyk, Systems Research Institute, Polish Academy of Sciences,
Warsaw, Poland

W0235361

SpringerBriefs in Computational Intelligence are a series of slim high-quality publications encompassing the entire spectrum of Computational Intelligence. Featuring compact volumes of 50 to 125 pages (approximately 20,000–45,000 words), Briefs are shorter than a conventional book but longer than a journal article. Thus Briefs serve as timely, concise tools for students, researchers, and professionals.

Paolo Massimo Buscema · Weldon A. Lodwick ·
Giulia Massini · Pier Luigi Sacco ·
Masoud Asadi-Zeydabadi · Francis Newman ·
Riccardo Petritoli · Marco Breda

AI: A Broad and a Different Perspective

 Springer

Author
See next page

ISSN 2191-530X   ISSN 2191-5318 (electronic)
SpringerBriefs in Applied Sciences and Technology
ISSN 2625-3704   ISSN 2625-3712 (electronic)
SpringerBriefs in Computational Intelligence
ISBN 978-3-031-80599-8   ISBN 978-3-031-80600-1 (eBook)
https://doi.org/10.1007/978-3-031-80600-1

This Springer imprint is published by the registered company Springer Nature Switzerland AG
The registered company address is: Gewerbestrasse 11, 6330 Cham, Switzerland

If disposing of this product, please recycle the paper.

Paolo Massimo Buscema
Semeion Research Center of Sciences
of Communication
Rome, Italy

Department of Mathematical and Statistical
Sciences
University of Colorado Denver
Denver, CO, USA

Department of Engineering
International Telematic University
UniNettuno
Rome, Italy

Giulia Massini
Semeion Research Center of Sciences
of Communication
Rome, Italy

Masoud Asadi-Zeydabadi
Department of Physics
University of Colorado Denver
Denver, CO, USA

Riccardo Petritoli
Semeion Research Center of Sciences
of Communication
Rome, Italy

Weldon A. Lodwick
Department of Mathematical and Statistical
Sciences
University of Colorado Denver
Denver, CO, USA

Pier Luigi Sacco
Department of Neurosciences,
Imaging and Clinical Sciences
University of Chieti-Pescara
Chieti, Italy

metaLAB (at) Harvard
Cambridge, MA, USA

Francis Newman
Department of Radiation Oncology
University of Colorado Denver
Denver, CO, USA

Marco Breda
Semeion Research Center of Sciences
of Communication
Rome, Italy

University Laboratory on AI
International Telematic University
UniNettuno
Rome, Italy

Preface

Artificial Intelligence aims to make thinking automatic. We have learned to make material procedures automatic: the engine, the refrigerator, the laser, the physical and chemical chains of matter and energy. But thought is different: thought does not seem to have mass, it is invisible. Yet even without mass, thoughts attract each other. Thought seems to be a word born in the plural.[1]

This is what Artificial Intelligence aims at: capturing the hard core of what completes the world of matter. The brain, in fact, appears to us as the material basis of thought and its main activity is to complete what matter lacks: we see a landscape because our mind has transformed a set of light impulses into an image. If that landscape is a little different from yesterday, our mind focuses on those parts of the landscape that have changed, as if it has realized that it has completed them incorrectly.

So, our mind is structured to add information to what matter shows us. The brain tries to complete every scene. The brain is a machine born to make predictions.

It is believed that thought arises spontaneously from local interactions between neurons in the brain. This is the point: massive physical and chemical interactions between parts of brain matter generate a global effect of a different nature, thoughts. Those thoughts that complete the very matter that produces them, show a natural predictive competence in the organism in which they emerge: if I think, I add, therefore I foresee.

The brain has the ability to predict. A brain is also present in insects, but their synapses are fixed from the beginning and, therefore, experience does not change them. Insects do not learn. So, in accordance with what was said earlier, they do not think. Yet the behavior of insects, especially when observed in their colonies, they appear to have thought.

Thinking seems to be linked to an organism's ability to form an internal representation of the context in which it is immersed. By manipulating these representations, the organism makes continuous analogies, that is, predictions about the environment

---

[1] This paper was published for the first time in an Italian Newspaper (copyright by Sole 24 Ore, section "Nova", 2018-09-24).

(cause-effect relationships, possible developments, the origin of certain signals, and so on). Using Artificial Intelligence, we create artificial neural networks able to represent the environment in which we immerse the data through many numerical vectors (layers). When two vectors of numbers that were formed in two different moments, from different environmental signals, that are "similar", then a thought might emerge: those two signals are somehow related.

Thus, thought might be considered as being formed by internally encoding similarities and differences in signals from the environment. These similarities add information to what the environment shows, and when their consistency is checked, it is as if they complete the environment itself. They predict it.

When therefore a colony of ants, each of which is devoid of thoughts, shows an organized and functional behavior for its own survival, it means that those ants show similarities and differences and therefore they are linked together. Consequently, not the individual ant, but their colony is a collection of thoughts that follow each other in time.

Thus the behavior of an ant colony is the visible representation of an internal representation of the thoughts of "nature". At any given moment, it is as if nature were predicting the subsequent behavior of the entire colony as the environmental signals change.

The essence of doing: If human beings think, then nature also thinks. Human beings themselves are a thought of nature thinking its nature. One aim of artificial intelligence is therefore to imitate the work of nature. That is, artificial intelligence attempt to create "think structures" that think and which also think their thoughts. That is to say, they think their own thinking.

Explaining the laws of nature is the goal of science. Artificial intelligence deals with those laws of nature whose result is immaterial: understanding, thinking, learning to learn, imagining. Thinking is the essence of doing. This essence is an invisible doing that, like a bridge over the void, connects material actions and events. It could be argued that the elementary units of artificial intelligence are information, which are connected to each other, form thoughts, that are connected to matter to create actions, behavior and culture. Thus, artificial intelligence aims at a physics of thought.

Information is only created within a brain that receives material perturbations. Without a brain that functions as a receiver, every perturbation is mere noise. But information is something intangible and abstract. And at the same time, information always needs matter to be transmitted. A thought, as an oriented and dynamic network of information that is even more abstract and invisible than the information it connects. Also, a thought can only live within a brain that is capable of modifying its connections and/or combining them, that is, to pass from one thought to another. If this dynamic stops, the brain stops thinking, it is a dead brain.

At this stage a hybrid concept arises: learning. Hybrid because learning is the completely abstract, but only materially verifiable, result of a chain of thoughts and actions that have taken place over time. The purpose of learning is to cancel the time that was necessary to achieve it. I see a material form and in an instant, I understand that it is the face of a friend, a relative or an enemy. All the time that

I have spent learning these differences, trials, errors, and therefore approximations between thoughts and events, is wiped out in an instant. I have learned and therefore now I understand.

Learning to learn: Thinking is much more than learning from experience (data). It seems that the human brain does not just learn, but takes it to a higher level, it learns to learn. This last concept is crucial. It is not a process by which an organism learns, but rather a process of learning to learn.

It is not a process whereby an organism simply learns in a recursive chain as long as it likes. Today's deep learning instead moves as follows. A mechanism of chain abstractions begins so that the last abstractions capture the fundamental invariants of an experience. Deep learning, understood in this way, is not a 'meta-learning', but a series of interpretative filters through which each level interprets the previous one. It is as if we were convinced that the more a story is passed down orally from one subject to another, the more it will show its hidden truths. The final result will not be the 'hard core' of that tale, but the interpretative culture of its narrators, gossip included.

Real deep learning, on the contrary, should be real meta-learning. From each specific learning the organism abstracts strategies will help it to learn more effectively a new situation that is completely different from the first one. For example: what do tennis and chess have in common? All the strategic meta-knowledge that the tennis expert might be able to adapt to learning the logic of chess, the concept of the feint, the attack simulation, the long throw, the defensive retreat, the spin on the ball, and so forth. In short, there is the similarity of the game of tennis and chess. This means intelligent systems are able to capture from an experience those abstract rules that can generate useful rules for learning another experience partially or totally different from the first one.

Analogies between distant worlds: Learning to learn, thinking and hence thinking our own thoughts means to create analogies and transfers of method between apparently distant and completely unrelated worlds. To build, in short, 'impossible bridges' between different worlds.

Thought normally crosses these impossible bridges back and forth: from Walt Disney's creations to the many-worlds interpretation in quantum physics, from the metaphors that emerge spontaneously in local markets to the theory of multi verses or of different types of infinity in mathematics. The specifics of thought seem to be its extraordinary ability to move the impossible into reality.

Science has the task of unveiling the 'real'. But when science deals with the human mind, then its objective becomes almost paradoxical: to reveal the reality of a mechanism capable of creating the reality of unreality. Artificial intelligence has to measure itself against this goal: to automate a system that creates invisible, even impossible objects and then connects them in a way that is often just as impossible.

This goal is impossible for artificial intelligence when thinking of achieving it through a large artificial neural networks with many horizontal layers (deep learning) that aims to identify the few fundamental invariants of all experience (big data). Projects of this type can be able to build robust systems of "stimulus-numerical representation-response", effective on specific domains of knowledge where, instead of the analogy between separate worlds, the probabilistic co-occurrence between different stimuli dominates. In the best-case scenario, an adaptive parrot will be constructed.

The importance of biodiversity: Nature itself and the human brain (its chief administrator for certain functions) suggest another way. Nature explodes its complexity over time through the diversity of the objects that it makes interact with each other. One could even say that biological evolution is the story of competitive and cooperative negotiations between different objects. The more diverse the contracting parties are, the more, if one meeting point is found, the outcome will powerfully change the future of all the others. We are arguing that what seems 'stupid' at one time, may turn out to be a 'genius' solution a moment later. Biodiversity is therefore the golden reserve of nature's creativity and balance.

The human brain functions in a similar way. A large number of neurons, organized in small networks each of a few layers (6 at the most), differing from each other in topology, function and hierarchical position. The brain seems to be the reign of the structural and functional biodiversity of the neural networks that compose it. This diversity is the basis of its efficiency: different networks understand and transmit different information to other networks, which in their own specific way transmit other information, in a complex temporal and spatial synchronism. This concert of unconscious artists that has developed over time cannot be replaced by a single 'soloist' playing each instrument alone. It would be like expecting the complex rhythm of movements in a crowded oriental bazaar to be reproducible by a single army of soldiers marching in a goose step.

Thus, the fact that our brain is structured as a complex society of specific networks is a visible sign of how our invisible mind works. Indeed, it is evidence of how complexity emerges almost spontaneously from the synchronization of bio-differences. The diversity between two components squeezed into the same space-time is the first of all conflict. Through negotiation, mistakes and approximations and finally cooperation. Cooperation is also, and perhaps above all, unconscious. As each component specializes in one type of response, it is the global response capacity of the whole that becomes more effective in dealing with any unforeseen external perturbation.

If I am wrong, I exist: A new aphorism could be coined to summarize the logic of complexly evolving systems, "if I am wrong, I exist, and if I exist it is because I am different." The word 'wrong' in the previous sentence is not a mistake. Each component of a complex system, in order to contribute to the whole, must necessarily try out many behaviors. Only a few will turn out to be appropriate for the whole system, while most will be wrong. These errors are the salt of the adaptive systems. It is like telling me how you are wrong and I will tell you who you are. The dynamic identity of each small neural network in the brain is formed on the basis of its operating errors. If there are too many of them and they do not correct themselves over time, the network in question will tend to reorganize itself. Otherwise, if not, that network will specialize in providing the correct answers that it has selected during the 'learning' process.

Making human thinking automatic: It becomes clear that thinking involves learning and learning is achieved by minimizing the errors that are necessarily made over time. But different neural networks learn different data in different ways, so they make different mistakes and develop different skills. It is, therefore, the collective and hierarchical learning of these different neural networks, which have different abilities and different weaknesses, that is the strength of truly deep learning ('if I am wrong I exist').

The artificial intelligence of the near future will have to pass from the form of 'violent learning' (many filter-layers in succession, crushed by an externally imposed objective function) to the form of really deep learning, based on the unconscious co-operation of many small differently specialized neural networks (many different points of view, each with its own internal objectives on the many aspects of the same problem).

It is as if artificial intelligence itself had to learn to see the human brain as a huge colony of insects, which, instead of moving around, establish adaptive networks of information exchanges among themselves according to data coming from outside and inside their system. The human brain is one of the most incredible and impossible thoughts produced by nature, a thought that generates thoughts about its own origin. Artificial intelligence is the only science that aims to make the invisible world of human thought visible, material and automatic. Developing true artificial intelligence should not just be about producing new applications and lots of money, but about better understanding who we are and how strange we are. Perhaps we are a whim of nature. But this whim has turned into what nature lacks. It is a thought that is in great need of a future.

Rome, Italy	Paolo Massimo Buscema
Denver, USA	Weldon A. Lodwick
Rome, Italy	Giulia Massini
Chieti, Italy	Pier Luigi Sacco
Denver, USA	Masoud Asadi-Zeydabadi
Denver, USA	Francis Newman
Rome, Italy	Riccardo Petritoli
Rome, Italy	Marco Breda

Introduction

The six chapters composing the backbone of this small book might appear disconnected and disharmonious. However, this is not the case. Each chapter constitutes a response to the six myths currently linked to the scientific, technological, and market success of Artificial Intelligence.

Chapter 1, indeed, debunks the "fake news" portraying artificial intelligence as a plausible representation of the functioning of the human brain. These are still two entities with vastly different structures, functionalities, and effects. Within this chapter, several of these significant differences and divergences are showcased.

Chapter 2 addresses the myth of a singular type of Artificial Intelligence capable of satisfactorily simulating every human cognitive and pragmatic behavior (General Artificial Intelligence). Contrary to this notion, the argument presented is that there exist at least three types of Artificial Intelligence, differing fundamentally in their overarching purposes, even if many of the algorithms they employ are commonly shared. We believe it is the third type of Artificial Intelligence, which we present, the one less recognized by "markets," journalists, and even the academic public, that is destined to be the most useful for the progress of human knowledge.

Chapter 3 dismantles the myth of "Big Data" as a fundamental element for constructing serious and useful artificial intelligence. It demonstrates how even within a single image, specific Neural Networks can identify hidden medically crucial information, invisible to traditional Deep Neural Networks trained on millions upon millions of images. There exist Neural Networks trained on a solitary image, capable through such training of rendering visible to the human eye and to other algorithms what these two actors would never be able to infer. The same philosophy is presented when illustrating how from small data sets representing an event, such as an epidemic or terrorist attacks, it is possible to estimate the present, past, and future geography of that process using only the geographical coordinates of the places where that process manifested. Hence, utilizing "small data" and "thick data" (dense data), one can develop an artificial intelligence of great practical utility and vast theoretical prospects.

Chapter 4 highlights the limitations of "Large Deep Neural Networks," comprised of millions of nodes and billions of parameters to be trained. Through blind tests known to all experts in the field, it is demonstrated how many small Neural Networks coordinated in a dynamic and uncorrupted parliament can achieve predictive results that even the deepest Convolutional Neural Networks (CNN) cannot fathom. Essentially, it is shown that a democracy composed of numerous limited yet competent individuals in different domains is much more efficient and effective than a grand enlightened tyranny composed of a single large Neural Network. Thus, the suggestion is to integrate the vertical "depth" of large Neural Networks with the horizontal depth of algorithms with diverse topologies and mathematics, mimicking what seems to be nature's optimal strategy: biodiversity as a tool to resolve unforeseen and sudden changes.

Chapter 5 showcases the limitation of the myth of hyper-specialization of various nodes within a Neural Network during its training phase. The mainstream of Artificial Intelligence seems to emulate the 20th-century model of hyper-specialization as the key to scientific and technological progress. Each expert becomes more proficient in their field of knowledge the smaller their field of knowledge; thus, they attain the pinnacle of knowledge when the area of their knowledge approaches "nothingness." This chapter demonstrates, through technical examples and real-world applications, that Neural Networks whose nodes develop a holistic learning of the data they are trained on are much more intelligent, useful, and accurate in predictions compared to a classic Neural Network that specializes the knowledge of its nodes based on co frequency.

Chapter 6 deeply criticizes the current myth that to develop a "conscious" artificial intelligence, the first step is to endow current Neural Networks with attentional competencies ("Attention is all you need"). Our conjectures on this subject vary significantly but are not isolated: for a system to be conscious, it is fundamental that it is "alive," and for it to be "alive," it must be designed from the outset as structurally unstable. The following reasonings imagine how abstract information projected into our 3+1-dimensional world is the matter that makes living systems, just as 3-dimensional bodies project shadows into 2-dimensional space. However, this chapter deals not with experiments or theorems but rather with conjectures. Nevertheless, the idea that living systems are rather complicated machines is not even a conjecture but the oversimplification of a dangerous illusion.

Contents

Chapter 1
The Parallels Between Deep Neural Networks and Modularity Theories of Brain Function

The recent success of deep neural networks in modeling complex perceptual and cognitive functions has led to comparisons between these artificial networks and the biological neural networks that make up animal and human brains. In particular, the layered architecture of deep neural networks, with each layer specialized for extracting specific types of features from the input, has invited analogies to the modular organization of brain regions specialized for certain cognitive processes. However, modularity theories of brain function overemphasize specialized areas at the expense of distributed processing and interconnectivity. While deep neural networks reflect a modular abstraction of brain function, network neuroscience offers a more nuanced view of cognition arising from both specialized regions and distributed networks.

The layered architecture of deep neural networks mimics hypotheses about the hierarchical and modular organization of the cerebral cortex [11]. Early visual areas detect simple features like edges and contours, while higher areas encode complex configurations like shapes and objects. This feedforward, bottom-up processing stream matches the architecture of convolutional neural networks used in computer vision. Convolutional networks contain successive layers of filters that extract increasingly abstract visual features, resembling the hierarchy of visual processing in the brain [34]. Beyond vision, modular theories posit domain-specific regions for functions like language, memory, emotion, and executive control. This functional segregation into circumscribed modules matches the hyper-specialized subnets in an ensemble artificial neural network. Just as a deep convolutional network focuses on image classification, recurrent networks specialize in sequence processing, and so on. The brain is analogized to a collection of specialized networks working in parallel.

However, network neuroscience offers an alternative perspective, highlighting integrative processing via long-range connections across distributed brain regions [2]. While areas like the fusiform face area are engaged preferentially by faces, they still interact with other sensory and association regions. Face perception arises

P. M. Buscema et al., *AI: A Broad and a Different Perspective*,
SpringerBriefs in Computational Intelligence,
https://doi.org/10.1007/978-3-031-80600-1_1

from coordinated activity across a widespread network, not a single area [1]. Graph theory provides tools for characterizing these complex network topologies. Functional integration between modules may offer greater behavioral complexity and flexibility than strictly segregated processing [9]. Hyper-specialized neural network architectures may fail to capture these nonlinear dynamics.

Evidence also indicates significant recurrent and feedback processing in cortical networks, contrasting with the predominantly feedforward architectures of deep learning models. For instance, predictive coding theory proposes that higher cortical areas send predictions down the sensory hierarchy to suppress redundant input [3]. Hierarchical predictive coding networks implement more biologically plausible recurrent processing [19]. Other biological principles missing from modular neural networks include robustness, adaptivity, spontaneous activity, and oscillatory dynamics [35]. Deep learning overlooks the complex temporal structure of neural dynamics.

The apparent inconsistency between modular specialization and distributed integration is reconciled by conceiving of the brain as a nested hierarchy of networks [21]. Cognitive functions require both specialized processors and domain-general hubs that integrate information. Optimal brain organization balances functional segregation of local regions and global integration [27]. Characterizing these multi-scale network motifs remains an active research direction. Principles from network neuroscience could improve deep learning, such as layered networks with significant lateral connections or ensemble models with asynchronous communication between specialists.

In fact, collective intelligence architectures move towards this direction, forming networks of specialized agents that exchange information to jointly solve problems [8]. Future machine learning systems may resemble natural collective intelligences like bacterial colonies more than cerebral cortices. The modular, massively parallel architecture of deep learning stacks represents just one abstraction of biological principles. Network neuroscience provides a more nuanced paradigm for translating brain theory into artificial intelligence. AI has focused on emulating compartmentalized functional modules, but the brain's power emerges from its interconnectedness.

1.1 Human Learning Has Little to Do with Big Data: What About Machine Learning?

The brain demonstrates robust intelligence across a wide range of environments and tasks, learning complex conceptual representations from limited, sparse data. In contrast, deep neural networks rely on vast datasets for training, struggling to generalize beyond their narrow domains without massive amounts of labeled examples. The data hunger and brittleness of deep networks contrasts sharply with the flexible, generalizable nature of human learning. Recent advances in distributed and dynamic neural network architectures better capture these properties of biological networks.

Deep convolutional networks achieve state-of-the-art performance on perceptual tasks like image classification by extracting hierarchical feature representations via successive layers of filters [18]. However, these feedforward architectures differ fundamentally from the brain in requiring orders of magnitude more training data. For example, AlexNet was trained on 1.2 million labeled images from ImageNet to classify objects into 1000 categories [16]. In comparison, humans can learn to distinguish thousands of objects from just one or a few examples. It is implausible for human brains to receive millions of carefully curated training examples for each visual concept through evolution and development. Deep learning thus fails to explain the efficiency of human learning.

The data efficiency of human cognition depends on integrating prior knowledge and inferences, not just passive feature detection as in feedforward deep networks. For instance, people leverage intuitive theories of physics and psychology when interpreting scenes and actions [4, 13]. Knowledge about objects and their relations enables rapid generalization to new contexts. Deep learning struggles to make such inferential leaps without extensive retraining. Humans also harness compositionality, recursively combining basic concepts into complex representations [17]. In contrast, deep networks treat each training input independently, lacking built-in relational structure.

How do biological networks achieve their robustness and adaptability with limited data? Recurrent and lateral connections allow forward, backward, and parallel information flow, implementing dynamic processing loops akin to inference [24]. Attractor networks can rapidly encode associations after only brief exposure through fast Hebbian learning [28]. Generative models leverage Bayesian principles to integrate top-down priors and bottom-up signals [12]. Predictive coding uses feedback hierarchies to match predictions flowing downward against the actual sensory data [26]. Such recursive processing depends on both localized specialization and distributed interconnectivity.

Sparser, more distributed neural networks better capture these computational properties of biological systems [22]. For example, embedding networks can learn compositional representations from very few examples by encoding knowledge of relations between objects and concepts [31]. Interaction networks leverage graph structure and message passing between nodes to build relational inductive biases, supporting fast generalization. Multi-task networks with shared underlying representations inherit commonalities between tasks, reducing data needs compared to training separate models [7].

Biologically inspired architectures such as Liquid State Machines and Echo State Networks also display greater data efficiency and learning capabilities closer to human cognition [20]. Their reservoir computing approach depends on sparse recurrent connections and temporal dynamics rather than static deep hierarchies. Networks incorporating spiking neurons, synaptic plasticity, and other neuroscience principles will continue bridging the gap between AI and biological learning [30].

Therefore, human-like learning requires overcoming the data hunger of current deep learning approaches in favor of more distributed, flexible neural architectures incorporating prior constraints and relationships. The seamless generalization and transfer of people emerges from commonsense wiring of an intuitive physics and psychology, not big data. Matching these abilities will require embedding more inductive biases into our models of intelligence. Neuroscience theories like predictive coding and attractor dynamics already inspire more powerful AI algorithms. A renewed focus on the ecological efficiency and sparse, compositional representations of biological brains over massive statistical models provides a promising path for developing more human-like machine learning.

1.2   Toward a Neurodiverse Machine Learning: The Power of Small, Numerous, and Diverse Artificial Neural Networks

Biological systems like the brain exhibit incredible diversity, with specialized modules working together to produce flexible and intelligent behaviors. The mammalian neocortex consists of billions of neurons of various types, arranged into specialized regions handling distinct functions like vision, movement, language, etc. Within each region, microcircuits of distinct neuron types self-organize to process information [14].

This neurodiversity allows biological systems to flexibly adapt to changing environments and learn complex tasks [33]. Damage to one region can be compensated by others, a property called degeneracy [10]. Highly degenerate systems are robust, evolvable, and open-ended in their learning capabilities [6].

Standard deep learning methods like CNNs lack this diversity. All parts are trained end-to-end to optimize one cost function, usually via backpropagation. Modularity and specialization arise to serve this singular objective [23]. This produces brittle systems that fail catastrophically when conditions change [29].

Neurodiverse machine learning systems aim to capture the adaptability of biological systems by similarly incorporating diversity and specialization. Rather than using one monolithic model, many small, specialized modules are trained independently on the same data before being integrated.

A general neurodiverse learning framework contains the following components:

- Ensemble of Diverse Learners

A collection of machine learning models with different architectures (ANNs, decision trees, SVMs, etc.) forms the ensemble [25]. Models should be small and constrained, to promote specialization. Each model is trained separately on the full dataset.

- Feature Generation

Raw data is passed through each model to generate new abstract features capturing its specific representation [32]. For ANNs, the last hidden layer output can be used as features.

- Data Rewriting

Original data is rewritten using the new feature sets from each model. This generates new data sets capturing diverse perspectives.

- Specialized Learning

More constrained learners are trained on each rewritten dataset. Restricted network architectures prevent overfitting to any one representation. Specialization arises naturally due to the limited input features.

- Fusion and Decision Making

A meta-learner combines the outputs of all models on a new data point to make the final prediction. Its architecture ensures that consensus between diverse models is needed for confidence.

The meta-learner plays a key role in neurodiverse learning. Two possible architectures are:

- Mixture of Experts

In the mixture of experts architecture [15], each expert ANN specializes on a partition of the input space. A gating network assigns weights to experts based on the input. These are used to make a weighted combination of expert outputs.

- Meta Neural Network

The meta neural network introduced in this manuscript is an unsupervised meta-learner [5]. Its input layer concatenates the outputs of all expert models. Connection weights to output units are calculated from confusion matrices and do not require true labels. Consensus between diverse models is needed for high confidence.

Compared to conventional deep neural networks, neurodiverse learning offers several advantages:

- Modularity and specialization for inherent robustness. Failure of one component does not break the system.
- Flexibility and adaptability to handle changing conditions. Models can be added or removed modularly.
- Interpretability arising from meta-learner analysis. Consensus and dissent between models is visible.
- Generalization from fused perspectives rather than overfitting to one.
- Compact expert models require less data than large networks. Knowledge distillation transfers benefits.

However, designing the components and meta-learner requires more craftsmanship than standard networks. Testing different configurations is also required for optimal performance.

There remain many open questions around optimally designing neurodiverse learning systems:

- How to select small neural architectures to promote specialization while avoiding underfitting?
- What algorithms beyond backpropagation should be included for maximal diversity?
- How to determine an optimal fusion mechanism for a given problem?
- Can competitive or cooperative interactions between models produce additional benefits?

Evolutionary approaches may help solve these issues, allowing neurodiverse systems to autonomously self-organize. Online learning scenarios where new data continuously arrives also favor such flexible, adaptable architectures.

As machine learning tackles increasingly open-ended, complex domains like robotics, scientific discovery, and language understanding, neurodiverse systems may prove essential. Their blend of specialization, diversity, and integration mirrors how biological systems achieve adaptive intelligence. While challenging to design, the promise of flexible, general artificial intelligence motivates continued research into this approach.

References

1. Apps MAJ, Tsakiris M (2014) The free-energy self: a predictive coding account of self-recognition. Neurosci Biobehav Rev 41:85–97. https://doi.org/10.1016/j.neubiorev.2013.01.029
2. Bassett DS, Sporns O (2017) Network neuroscience. Nat Neurosci 20(3):353–364. https://doi.org/10.1038/nn.4502
3. Bastos AM, Usrey WM, Adams RA, Mangun GR, Fries P, Friston KJ (2012) Canonical microcircuits for predictive coding. Neuron 76(4):695–711. https://doi.org/10.1016/j.neuron.2012.10.038
4. Battaglia PW, Hamrick JB, Tenenbaum JB (2013) Simulation as an engine of physical scene understanding. Proc Natl Acad Sci 110(45):18327–18332
5. Buscema PM (1998) Meta Net: the theory of independent judges. Subst Use Misuse 33(2):439–461. https://doi.org/10.3109/10826089809056240
6. Buscema M, Sacco PL (2017) Digging deeper on "deep" learning: a computational ecology approach. Behav Brain Sci 40:e256
7. Caruana R (1997) Multitask learning. Mach Learn 28(1):41–75
8. Dafoe A, Hughes E, Bachrach Y, Collins T, McKee KR, Leibo JZ, Larson K, Graepel T (2020) Open problems in cooperative AI. arXiv preprint arXiv:2012.08630
9. Deco G, Tononi G, Boly M, Kringelbach ML (2015) Rethinking segregation and integration: contributions of whole-brain modelling. Nat Rev Neurosci 16(7):430–439. https://doi.org/10.1038/nrn3963
10. Edelman GM, Gally JA (2001) Degeneracy and complexity in biological systems. Proc Natl Acad Sci 98(24):13763–13768. https://doi.org/10.1073/pnas.231499798
11. Felleman DJ, Van Essen DC (1991) Distributed hierarchical processing in the primate cerebral cortex. Cerebral cortex (New York, N.Y.: 1991) 1(1):1–47
12. Friston K, Kiebel S (2009) Predictive coding under the free-energy principle. Philoso Trans Royal Soc B: Biol Sci 364(1521):1211–1221
13. Gerstenberg T, Goodman ND, Lagnado DA, Tenenbaum JB (2015) How, whether, why: causal judgments as counterfactual contrasts. In: Proceedings of the 37th annual conference of the cognitive science society (CogSci 2015). Cognitive Sci Soc 782–787

14. Harris KD, Mrsic-Flogel TD (2013) Cortical connectivity and sensory coding. Nature 503(7474):51–58. https://doi.org/10.1038/nature12654
15. Jacobs RA, Jordan MI, Nowlan SJ, Hinton GE (1991) Adaptive mixtures of local experts. Neural Comput 3(1):79–87. https://doi.org/10.1162/neco.1991.3.1.79
16. Krizhevsky A, Sutskever I, Hinton GE (2012) Imagenet classification with deep convolutional neural networks. Adv Neural Inf Proc Syst 25
17. Lake BM, Ullman TD, Tenenbaum JB, Gershman SJ (2017) Building machines that learn and think like people. Behav Brain Sci 40:e253
18. LeCun Y, Bengio Y, Hinton G (2015) Deep learning. Nature 521(7553):436–444
19. Lotter W, Kreiman G, Cox D (2016) Deep predictive coding networks for video prediction and unsupervised learning. arXiv preprint arXiv:1605.08104
20. Lukoševičius M, Jaeger H (2009) Reservoir computing approaches to recurrent neural network training. Comput Sci Rev 3(3):127–149
21. Meunier D, Lambiotte R, Bullmore ET (2010) Modular and hierarchically modular organization of brain networks. Front Neurosci 4:200
22. Olshausen BA, Field DJ (2004) Sparse coding of sensory inputs. Curr Opin Neurobiol 14(4):481–487
23. Olshausen BA, Anderson CH, Van Essen DC (1993) A neurobiological model of visual attention and invariant pattern recognition based on dynamic routing of information. J Neurosci 13(11):4700–4719. https://doi.org/10.1523/JNEUROSCI.13-11-04700.1993
24. Orbán G, Berkes P, Fiser J, Lengyel M (2016) Neural variability and sampling-based probabilistic representations in the visual cortex. Neuron 92(2):530–543
25. Polikar R (2006) Ensemble based systems in decision making. IEEE Circ Syst Mag 6(3):21–45. https://doi.org/10.1109/MCAS.2006.1688199
26. Rao RP, Ballard DH (1999) Predictive coding in the visual cortex: a functional interpretation of some extra-classical receptive-field effects. Nat Neurosci 2(1):79–87
27. Shine JM, Poldrack RA (2018) Principles of dynamic network reconfiguration across diverse brain states. Neuroimage 180:396–405
28. Sprekeler H (2017) Functional consequences of inhibitory plasticity: homeostasis, the excitation-inhibition balance and beyond. Curr Opin Neurobiol 43:198–203
29. Szegedy C, Zaremba W, Sutskever I, Bruna J, Erhan D, Goodfellow I, Fergus R (2013) Intriguing properties of neural networks. arXiv preprint arXiv:1312.6199
30. Thalmeier D, Uhlmann M, Kappen HJ, Memmesheimer RM (2016) Learning universal computations with spikes. PLoS Comput Biol 12(6):e1004895
31. Tokmakov P, Wang YX, Hebert M (2019) Learning compositional representations for few-shot recognition. In: Proceedings of the IEEE/CVF international conference on computer vision, pp 6372–6381
32. Van der Maaten L, Postma E, Van den Herik J (2013) Dimensionality reduction: a comparative review. TiCC TR 2009-005
33. Whitacre JM, Bender A (2010) Degeneracy: a design principle for achieving robustness and evolvability. J Theor Biol 263(1):143–153. https://doi.org/10.1016/j.jtbi.2009.11.008
34. Yamins DL, Hong H, Cadieu CF, Solomon EA, Seibert D, DiCarlo JJ (2014) Performance-optimized hierarchical models predict neural responses in higher visual cortex. Proc Natl Acad Sci 111(23):8619–8624
35. Yuste R (2015) From the neuron doctrine to neural networks. Nat Rev Neurosci 16(8):487–497

Chapter 2
A Theoretical Introduction

This short monograph looks at artificial intelligence (AI) from a point of view that may be different to the current ones that exploit the computational power of current computers to analyze "big data" using many layers in a neural network in which nodes within receive and capture *localized* data from their input layer(s) and deliver these to their output layer(s). That is the emphasis is often on big data and computational power allowing for AI systems with a huge number of layers. Clearly, when analyzing all text that exist on the internet, such as CHAT GPT, big data and computational power with AI systems that have many layers are needed. However, the variety of problems involving AI involve other approaches.

This monograph makes a case that for many problems, small data is sufficient and that nodes can encode more global information yielding a more robust analysis. Moreover, in addition, often the implementation of a large set of different neural networks working in together with these new structures yields superior results for some problems. The components of this perspective, both their theoretical foundation as well as their application with comparisons to some currently used neural networks is presented. The use of evolutionary algorithms is key to approaches that are able to adapt to changing environments and so key AI.

The idea of simulating the human mind on a computer emerged around the 1950s. This idea turned into a project known as Artificial Intelligence (AI). The early attempts highlighted the difficulty of this task requiring significant human and financial efforts and resulting in many failures [1, 2].

Today after 70 years of attempts, we seem to be close to the goal of constructing algorithms capable of simulating human thinking to the extent of easily deceiving another human about their true human or artificial identity. It is believed that we are on the brink of creating beings like ourselves [3, 4].

The social, economic, mechanical (for example, autonomous vehicles), and medical implementations of these recent algorithms, particularly deep neural networks, is evident [5]. When used for "good" they can assist us in various ways. But if used for "evil" or the "wrong", they can be destructive. Many "intellectual"

© The Author(s), under exclusive license to Springer Nature Switzerland AG 2025
P. M. Buscema et al., *AI: A Broad and a Different Perspective*,
SpringerBriefs in Computational Intelligence,
https://doi.org/10.1007/978-3-031-80600-1_2

jobs can be automated while new types of "human" work are ready to emerge. In fact, who else but human subjects, understanding how these algorithms work, will control, correct, and train artificial intelligence at work?

The exponential progress of AI in the last 20 years is based on some easily listed theoretical and experimental assumptions:

a. The human brain is a complex machine capable of producing emergent effects such as consciousness and self-awareness. Despite its complexity, the human brain can operate as a machine and, therefore, can be simulated and reproduced [6].

b. The human brain does not follow fixed and pre-established rles. It is a machine that creates its local rules by functioning and interacting with itself and the "world" and adapting to both. Thus, building algorithms with numerous operating rules is not always useful. It would be like representing all the information in a movie with a single photograph. It is better to construct a semi-empty structure initially, capable of self-modification over time in relation to the data it experiences. The only rules it will have are the constraints of its functioning and the objectives it must achieve each time (which can also be learned) in order to learn autonomously from the data it encounters or receives [7]. Learning from making mistakes and making decisions to make "improvements" are also important parts of human consciousness and so also for algorithms.

c. These intelligent algorithms become intelligent by interacting with data (experiences) and so they need training [8]. However, for them to function acceptably they must be trained not with just a few data points but often with massive amounts of data (big data). Without these huge amounts of data, intelligent algorithms may understand very little and make disastrous errors. An intelligent algorithm should be like a newborn, who after seeing 3 or 4 cats, perhaps can recognize dozens of cats it has never seen before. Something different must be happening in human processing of data.

d. The algorithms that have proven most effective in this endeavor are called *deep neural networks* (DNNs: [5, 9–11]). They can be very large in size receiving a multitude of inputs often in parallel especially for image processing and understanding. These networks are also vertically extensive with many layers of artificial neurons between input and output each interpreting the previous state. At times such a processing of many layers is like the game where a story is whispered secretly in the ear of the next person in line who then tells it to the next, and so on. The last person may end up with a completely different story from the original one. The cycle is then repeated with appropriate corrections. However, the questions arise. Who decides what to correct? How much to correct? In what way is it to be corrected? Nevertheless, what seems to matter is that in the end, the algorithm tells a story resembling a narrative [12].

The problem is that these assumptions may be false when compared to the fundamental goal of creating algorithms that function similarly to the human mind. It is not certain that the human brain is a machine. On the contrary, machines are combinations of more or less elementary parts. The human brain (and not just it) starts from a cell, which divides over time, first into two cells. then 4. 8. 16. 32. 64. 128. 256. 512. 1024. 2048. 4096. and so on. generating billions of cells (around 4×10^{13}), each of which has developed a structure and functions different from the others depending on where it was born or migrated. Cells seem to behave more like past human populations migrating and adapting to new environments rather than like pieces of a complicated airplane that are defined from the beginning and suffer consequences if they move. In other words, any machine (including software) does not grow over time and its components do not change their relationships nor their structure over time. They remain fixed. Neurons work differently.

2.1 Approaching the Subject

When studying a natural or cultural system whose functioning appears highly complex the current scientific method offers two possible strategies:

1. ***The top-down approach***: The top-down approach defines many rules and rules of rules that in parallel and sequence simulate the behavior of the system in question on a computer. Then this simulation is statistically checked in a sophisticated manner to guarantee its correctness. This is the traditional scientific method. However, this approach only works if the system being studied has a finite number of rules and meta-rules and follows them. Therefore, this scientific procedure gives results for systems that are "tautological" in meaning so that all the information they need is already contained in the hundreds of thousands of elementary pieces from which they were built. These systems can be described as "**complicated systems**". That is, combining their elementary units in all possible ways (the number of permutations is given by the factorial of the number of elements in the system) will yield at least one configuration that works as observed before disassembling it. Another characteristic of complicated systems is that the longer they function the worse they perform. Time does not make them grow and improve. Instead, time deteriorates them and they require maintenance from human subjects to reduce the accumulated "noise" during their lifecycles. This is a natural increase in entropy (amount of disorder) in an isolated system.

2. ***The bottom-up approach***: This second approach is useful for systems that create and update rules while functioning and, therefore, generate new information during their lifecycle. Imagine a living being growing over time from its first cell to the final epigenetic landscape we call a human being. This procedure is called the "Bottom-Up" approach, which is opposed to the previous one. Such systems are "**complex systems**" as they increase the quantity and quality of the information they produce over time and increase their internal order (reducing their entropy) like the human brain that selects over time the most useful connections for its functioning. Complex systems are, therefore, adaptive systems that spontaneously organize based on the local interactions of their elements. These

local interactions give rise to a global behavior that bears no linear relationship to the sum of the individual local behaviors of the same complex system. Additionally, complex systems cannot be disassembled into their "pieces" like a car and then reassembled. Each "piece" of a complex system seems to function only as long as it "feels" it belongs to the global system of which it is a part. This type of system can only be analyzed through the Bottom-Up approach, by interacting with its components until a new and global order emerges spontaneously from these local interactions. The "scientific gesture" required is the opposite of reductionism. Once the basic elements of a complex system are isolated, they must interact through appropriate equations and data to reassemble into the system from which the analysis started. In other words. they must self-reassemble autonomously. If this happens, it means that something is understood about their functioning. The current AI uses this strategy while classical AI known as "expert systems" used the first approach. It is undeniable that the current AI has made the decisive leap towards a new science in recent years.

2.2 Types of AI

The question that arises at this point is: What are the different types of Artificial Intelligence systems? Empirically, there are infinite possibilities, but from a theoretical perspective, we mention three types of AI systems that exist and are distinguished based on their primary objectives.

1. **The Simulative Approach**: The objective of this AI type is to understand how the human brain works. It involves writing software and building hardware to simulate some brain functions and then compare the behavior of the software and hardware with what is known about the human brain. If the simulation produces satisfactory results, this means that the AI architecture and its learning laws bear interesting similarities to how the brain functions. This allows new hypotheses to be made, new simulations to be constructed, and explanations for the functioning of the yet poorly understood brain process etc. In short the simulative approach aims to increase our knowledge of the functioning of the human brain and its potential degenerations. AI serves as a tool to achieve this objective.

2. **The Emulative Approach**: The objective of this second approach is to build a machine that exhibits the same performance as a human subject to the point of deceiving them about its mechanistic nature. Moreover, the goal is to construct machines that outperform humans in terms of speed, information precision, memory capacity, and the ability to make analogies and inferences. To achieve this objective, it is not necessary for the machine to generate its output in the same way a human would. The "way" is not a constraint as it is in the simulative approach. The quality of the result, its informational completeness, and its execution speed are the parameters to evaluate its performance. If it surpasses any human subject, the fact that it operates differently might be an advantage in terms of knowledge. An example of this type of AI is the Convolutional Neural

Networks (CNNs), one of the most famous deep learning algorithms used for image analysis. CNNs learn through a process called "error backpropagation", which is one of the main algorithms used in deep learning. However, there is nothing similar to error backpropagation in the human brain. Human learning is likely a combination of various mechanisms including synaptic plasticity (the ability of synapses to change their strength in response to neural activity), structural brain remodeling, synaptic pruning (elimination of less-used synapses), and other forms of neural adaptation. Moreover. the human brain is extremely complex and the learning process involves a vast network of interconnected brain regions each playing a specific role in visual and other sensory information processing. Furthermore, during learning, CNNs can change their excitatory synapses to inhibitory and vice versa depending on the type of images they need to learn. Nothing similar occurs in the human brain where inhibitory synaptic weights are not modifiable through a learning process as they are in artificial neural networks. When we talk about "inhibitory weights" in the human brain, we refer more to a physiological property based on the density or properties of inhibitory synapses formed by neurons. These inhibitory synapses play a fundamental role in balancing neural activity and modulating the responses of neural networks during information processing. However, the "inhibitory weights" in the human brain are not changed through a learning process. Despite these significant differences the emulative approach often achieves objectives comparable to those of the human brain in image analysis. The human brain can inspire the emulative approach, but the way the latter works is not a constraint for this approach.

3. **The Investigative Approach**: This approach differs from the previous two. It is inspired by the human brain as a system of considerable complexity, but its objective is neither to simulate nor to emulate it. The primary goal of the physical approach is to understand the invariant laws governing complex systems, the laws through which individual behaviors organize into collective behaviors which often bear little resemblance to the individual behaviors at the outset. Examples include how individual atoms aggregate into molecules, how molecules form proteins, how the behaviors of individual ants organize into colonies, how individual bees organize into structured swarms, how individual starlings aggregate into dynamic and complex flight patterns that are inexplicable based on their individual behaviors, how continuous dynamic processes on earth transform into sudden earthquakes of great magnitude, and so on. In short, how is it that simple natural structures transform into unusually complex processes both natural and cultural. In other words how life emerges and how more complex life forms evolve from elementary life forms. Furthermore, how currently considered unpredictable phenomena form. To achieve this objective the bottom-up approach is indispensable. As this objective is currently implemented, it usually does not lead to the formation of complex systems and can be described through "closed equations". The focus should be on constructing algorithms that behave similarly to the "natural algorithms". Governing these (natural) algorithms is learning over time and how these systems create local rules, adjust them, and

change them. Finally, how these systems, during this process, exhibit behaviors that seem unpredictably complex from classical perspectives. For this reason, the algorithms to be designed are not "models" of the reality being studied but rather they are "meta-models" that interact with real data to spontaneously generate the subjective model constructed by the specific reality under study. Deep and shallow artificial neural networks (ANNs) are part of the toolkit of adaptive algorithms useful for this purpose. However, not only the ANNs known in the literature but also new types of adaptive algorithms should be researched using not just a single learning law (like gradient descent even with additional flavors). They should include new laws of learning and artificial evolution. The advantages of this approach over the others are many. It does not attempt to build a "wise parrot" that operates as a universal statistical approximator of human languages (for example, Chat GPT and similar systems). Instead, it extracts all available information, especially hidden information not visible to other algorithms from each data set representing a piece of reality. It is the capturing of hidden information in data that allows the investigative approach to present itself as useful in the development of the next phase of AI. In fact, the algorithms designed and to be designed for this purpose will also be useful for the emulative and the simulative approaches. For example, if we provide Chat GPT, an example of emulative AI, with all the law enforcement documents on a murder case, the software will be able to provide an excellent summary of the case and answer questions if the answers are present in those documents. However, it will not be able to make any sensible hypothesis about the culprit. The reason is simple. It was not built for that purpose. In this case. Chat GPT must pass the data it possesses in the form of a structured data set to another set of algorithms specialized in capturing hidden information in data (examples of physical AI). These algorithms, once they have completed their work will return the results of their investigation to Chat GPT which will then explain them in natural language to investigators. This collaboration between different artificial organisms should become a general framework for the new AI. This new system would be like Chat GPT but should act as a moderator of a TV program interpreting questions from the public on any topic and passing them on to experts for analysis according to their expertise. Then the moderator reformulates them in a way understandable to a general audience. This could be an interesting artificial collaboration.

References

1. Haenlein M, Kaplan A (2019) A brief history of artificial intelligence: on the past, present, and future of artificial intelligence. Calif Manage Rev 61(4):5–14. https://doi.org/10.1177/000812 5619864925
2. H. Jaakkola, J. Henno, J. Mäkelä and B. Thalheim (2019) Artificial intelligence yesterday, today and tomorrow. In: 2019 42nd International convention on information and communication

technology, electronics and microelectronics (MIPRO), Opatija, Croatia, 2019, pp 860–867. https://doi.org/10.23919/MIPRO.2019.8756913

3. Scarfe P, Watcham K, Clarke ADF, Roesch EB (2023) A real-world test of artificial intelligence infiltration of a university examinations system: a "Turing test" case study. https://doi.org/10.31234/osf.io/n854h

4. Dodig-Crnkovic G (2023) How GPT realizes Leibniz's dream and passes the Turing Test without being conscious. Comput Sci Math Forum 8(1):66. https://doi.org/10.3390/cmsf2023008066

5. Yan WQ (2023) Computational methods for deep learning: theory, algorithms, and implementations, 2nd edn. https://cerv.aut.ac.nz/wp-content/uploads/2023/06/Deep_Learning_Book.pdf

6. Churchland PS, Sejnowski TJ (1992) The computational brain. MIT

7. Arbib MA (2003) Handbook of brain theory and neural networks. MIT

8. Ma J, Wang J, Han Y, Dong S, Yin L, Xiao Y (2023) Towards data-driven modeling for complex contact phenomena via self-optimized artificial neural network methodology. Mech Mach Theory 182

9. Buscema M (1998) Theory: foundations of artificial neural networks, in substance use & misuse. Marcel Dekker Inc., New York, pp 17–164. 33(1) (theory)

10. Bengio Y (2009) Learning deep architectures for AI. Mach Learn 2(1):1–127

11. Schmidhuber J (2015) Deep learning in neural networks: an overview. Neural Netw 61(2015):85–117

12. Goodfellow I, Bengio Y, Courville A (2016) Deep learning. MIT Press, Cambridge (MA), p 2016

Chapter 3
The Myth of Big Data

The acceleration of AI in the last 20 years is linked to the ability to process large amounts of data (big data) and the increase in machine computational capacity and speed achieved through the re-optimization of multiple graphics processing units (GPUs) originally used for video games. The mathematical ideas underlying this evolutionary leap are few that are new and many that are classical. These are the backpropagation principle extended to multiple layers of intermediate units between the input and output (1970–1980) and the use of convolutional filters for ANNs (1980s). ANNs that use backpropagation use a few new tricks to allow a multi-layered ANN to learn. Throughout these new developments is big data that has proven to be fundamental. However, some questions naturally arise.

We know that the Central Limit Theorem informs us that if you take sufficiently large samples from a population, the samples' means will be normally distributed even if the population isn't normally distributed. Consequently, the tails of this distribution tend to zero. In other words, it is possible to construct samples that represent a problem without having to collect all the data that manifests that problem. Based on this theorem, statisticians can analyze voting patterns with small samples of voters with great and demonstrated precision without having to wait for the counting of all the votes contained in the ballot boxes. Therefore, known statistical mathematics and the resulting practical applications seem to suggest that in data analysis enormous amounts of data are not always necessary. What is needed, however, is a few dense data points meaning data whose variety and relative quantity statistically represent the problem being analyzed. Hence, of the three "Vs" that characterize big data (volume, variety, and velocity of storage), perhaps only two are needed—variety and velocity.

This insight leads one to ask the following questions. Are big data necessary for deep ANN learning? Are they useful for data commerce and monopolization? Big data may be, in many cases, fat data that needs substantial cleaning before being used to train one or more ANNs to avoid them learning nonexistent connections and making deadly mistakes. The human brain seems to rapidly acquire dense and highly

© The Author(s), under exclusive license to Springer Nature Switzerland AG 2025   17
P. M. Buscema et al., *AI: A Broad and a Different Perspective*,
SpringerBriefs in Computational Intelligence,
https://doi.org/10.1007/978-3-031-80600-1_3

diverse data to perform its inferences and analogies. It does not appear interested in or capable of storing enormous volumes of data. In short the human brain seems more oriented toward learning useful information from the small painting of the "Mona Lisa" rather than giants undergoing a steroid treatment".

3.1 Small Data and Dense Data: One Image at a Time

First, the issue of big data is analyzed via some examples of what is being advocated. In traditional image analysis with deep ANNs (CNNs) millions of images and billions of hyperparameters are often used. However, how much hidden information, even to the expert eye, is embedded in a single medical image? We now demonstrate how it is possible to train a specific ANN (Active Connection Matrix—ACM) to learn the hidden information in a single image [1–6]. To achieve this, however, the classic approach through which current ANNs are trained must be modified in the following way.

- First and foremost. a change in perspective is necessary. It is not I who filters the image through a convolutional kernel, but rather each individual pixel of the image "speaks" from its point of view to the pixels around it deciding how much to activate and how to modify the activations of its neighboring pixels. It is a shift from an Aristotelian view of the image. where the kernel decides what is relevant to a Socratic view where all pixels collectively look "within", engage in "dialogue" with each other, and elicit their intrinsic model.
- For this to happen, each pixel of the image must be connected to its immediate neighbors through independent bidirectional connections that dynamically change through unsupervised "reinforcement learning" laws. It is through this collective dialogue that the starting pixels of the image begins to change until they reach a collective agreement, via the so-called minimization of the "loss function" through reinforcements and inhibitions cycle after cycle.
- In this new form of learning the pattern to be learned is no longer the one image itself (which is just one), but rather each pixel of the image with its immediate neighbors. The entire image therefore, transforms into a matrix of active and adaptive connections among the pixels that make up the image. This is why the Active Connection Machine (ACM) algorithm is useful for this type of analysis. Thus, the image itself transforms into an ANN where each of its nodes (pixels) is different from all others in position and in the neighboring pixels that act as its context. Such ANNs are not composed of one or more layers. Each pixel is a singular node of this new ANN equipped with its incoming and outgoing weights (connections) as well as with its immediate neighbors. It would be appropriate to define these ANNs as neural structures with a pertinent geography.
- The learning and evolution laws of these ANNs can be diverse and can become an autonomous exploratory field. This means that this type of ANN can also learn multiple images together to intelligently merge them. They can also rewrite each

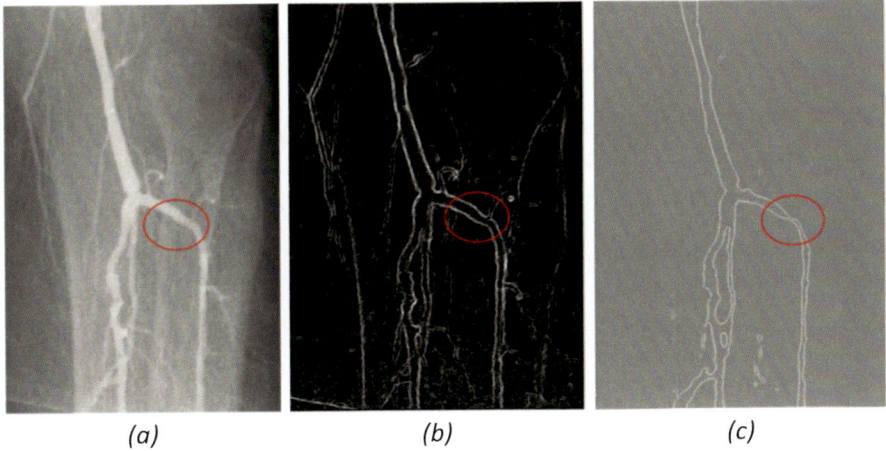

(a)   *(b)*   *(c)*

Fig. 3.1 a Original image; **b** original image processed by Sobel Filter; **c** original image processed by ACM system

image of a data set in a very different way and then process all the transformed images through a standard CNN.

3.1.1 The Hidden Stenosis

This process of taking an image and transforming it into an ANN for data image understanding is illustrated next. Figure 3.1a represents a frame of a subtractive angiography with contrast medium of a popliteal artery. This image does not show any particular problems to the expert eye of several radiologists. In Fig. 3.1b the same image is passed through a traditional filter offering the same, if not fewer, pieces of information compared to the original. In Fig. 3.1c, the same image is processed through an ACM network for 30 cycles [2]. This image appears very different from the original and highlights a stenosis (red circle) that the original image did not show. After a new angiographic examination, the stenosis was confirmed. Moreover, invisible stenoses were also verified in dozens of other experiments on patients' arteries.

3.1.2 The True Lumen of the Artery

The medical field has another issue that arises in the quantitative angiography of arteries: What is the real lumen of an artery. To determine this, contrast medium is used which is a compound based on iodine that is opaque to X-rays, as highlighted by the angiogram. However, what is visualized with the contrast medium is not the

Fig. 3.2 a, b On the left. The lumen of the artery according to quantitative angiography (lumen in black). And on the right, the lumen of the same artery measured using an ACM neural network (lumen in red)

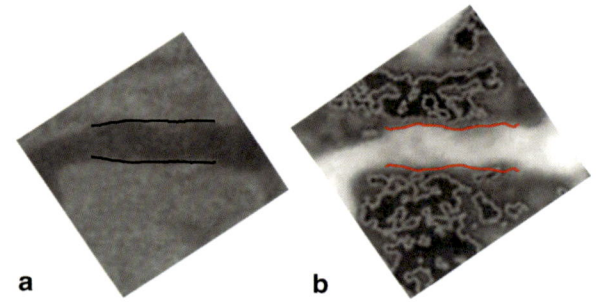

a   b

actual blood flow inside the studied artery but rather the flow of the contrast medium itself. We assume that the two flows are equivalent but due to the different density and composition of these two substances we cannot be certain. Therefore, it is not certain that the lumen of an artery measured through the flow of the contrast medium is the same lumen as the one through which the blood flows. When an artery is large enough, it is possible, however, to measure its lumen with good accuracy using IVUS (Intravascular Ultrasound). a diagnostic procedure based on a catheter used to visualize the interior of a coronary artery providing real-time imaging. IVUS is an invasive procedure that can only work with large diameter arteries. In the following example IVUS will be used as the gold standard to verify and compare the accuracy of measuring the lumen of different arteries through quantitative angiography with contrast medium as opposed to the lumen measured by an ACM neural network. These analyses and measurements have been carried out by researchers at Semeion Research Center in Rome and at the Monzino Cardiology Center in Milan[1] [7]. Figure 3.2a shows the lumen of an artery through quantitative angiography (in black) while in Fig. 3.2b the lumen of the same artery is measured using an ACM neural network (in red).

Figure 3.3a–c show the comparison between these two lumen hypotheses and the real lumen measured through IVUS (in green) on different arteries.

Figure 3.4 shows the results of various measurements conducted by comparing the angiography system with the ACM neural networks.

These are some examples of the investigative approach of AI that are added to the more standard AI approaches, highlighting hidden information in small but dense data sets, in this case. A single image.

---

[1] Semeion: Paolo Massimo Buscema; Monzino: Damiano Baldassarre, Mauro Amato.

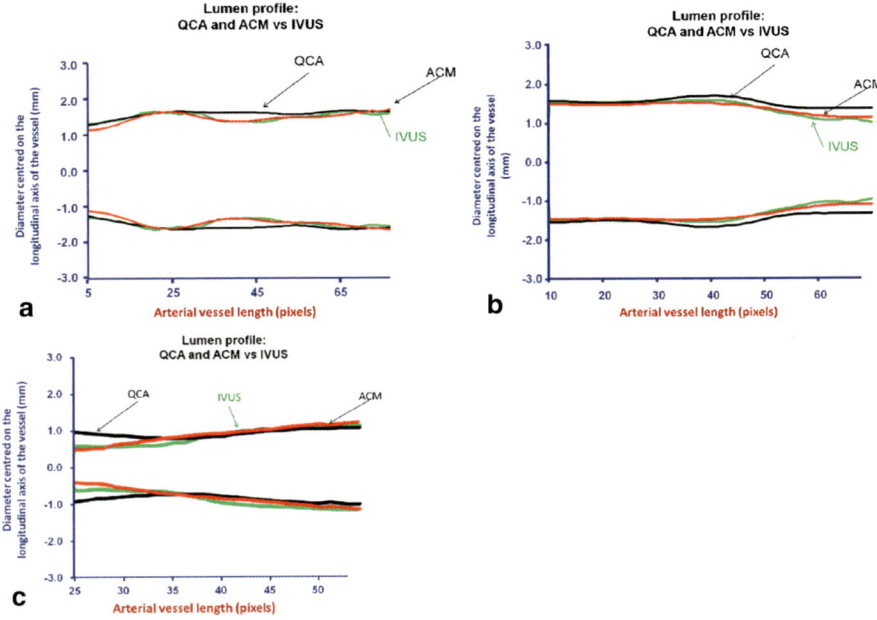

Fig. 3.3 a–c The lumen measured with IVUS (green) compared with the lumen measured with the angiography machine (black) and that measured with ACM (in red) in three different sections of the artery

Concordance between the diameter of the arterial lumen measured with angiography and with IVUS (gold standard) using conventional or reprocessed angiographic images with ACM algorithms

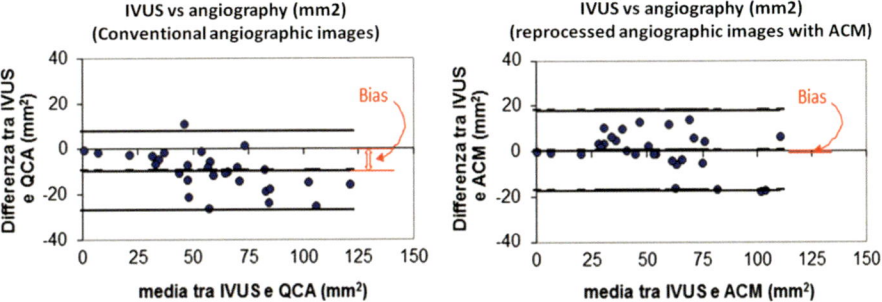

Fig. 3.4 The measurement of arterial lumen performed on images reprocessed with ACM algorithms is more accurate than that performed on conventional angiographic images

3.2 Small Data and Dense Data: Only Latitude and Longitude

Imagine an extremely small data set of N points in a two-dimensional space where only the latitude and longitude of each point are known. These points could represent the initial locations where a new epidemic has appeared or the positions of places where terrorist attacks have been carried out. In all cases let's assume that the chronological order of these events is unknown as well as their frequency at each location (that is, the number of cases of a certain epidemic in a particular city). It is a data set that is very scarce in information. If there were 13 points under examination (N = 13), then you would only be working with 26 numbers (13 latitudes and 13 longitudes). It may seem like a situation where the AI often portrayed in the media is unusable and futile. However, it is not so. One must always bear in mind that starting from 13 points in a two-dimensional space there are 239,500,800 different paths to visit all those points. passing through each of them only once and not excluding any of them from these paths, (N-1)/2 factorial. If the points were only 30 (60 numbers), then the number of different paths would be equal to 4,420,880,996,869,850,000,000,000,000,000 different itineraries.

3.2.1 Germany 2011: Escherichia Coli

Thus, everything that seems small is not actually small. And that's why if you have a "poor" data set that only indicates the geographic coordinates of the first 13 cities where an epidemic is occurring, using a novel geographic profiling system like the Topological Weighted Centroid (TWC) it is possible to obtain the possible hypotheses from these "few" data points as follows [8–15].
    ：

- Where this epidemic started.
- How it will spread in the short and medium term.
- Where it will tend to concentrate in the near future.

It's not magic but mathematics. The hypotheses that emerge are not certain truths but useful ones. Table 3.1 shows the original fax sent from Gottingen Center to University of Colorado at Denver of the first 13 cities where at least one case of Escherichia Coli was reported in Germany in May 2011, known as the vegetable epidemic.

The reading of Table 3.1 leads us to believe that since the city of Hamburg has the highest number of cases (Q = 59) Hamburg is likely to be the most probable focal city of this epidemic. This was also the belief of German researchers in May 2011. Now. let's imagine that our AI system called TWC, has only the latitude and longitude of the 13 cities from Table 3.1 as input without any other information (see Table 3.2).

Table 3.1 Original fax sent from Gottingen Center to University of Colorado at Denver of the first 13 cities with cases of Escherichia Coli up to May 2011

ID	State	City used	Why used	Lat	Long	Q
1	Hamburg	Hamburg	Exact match	53° 33′ 55″ N	10° 00′ 05″ E	59
2	Bremen	Bremen	Exact match	53° 4′ 33″ N	8° 48′ 27″ E	11
3	Schleswig–Holstein	Kiel	Capital	54° 19′ 31″ N	10° 8′ 26″ E	21
4	Mecklenburg-Vorpommem	Schwerin	Capital	53° 38′ 0″ N	11° 25′ 0″ E	10
5	Hesse	Frank flirt	Largest city	50° 6′ 37″ N	8° 40′ 56″ E	31
6	Saarland	Saarbrucken	Capital	49° 14′ 0″ N	7° 0′ 0″ E	5
7	Lower Saxony	Hanover	Capital	52° 22″ N	9° 43′ E	28
8	North Rhine-Weatphalia	Duesseldorf	Capital	51° 14′ N	6° 47′ E	31
9	Berlin	Berlin	Exact match	52° 30′ 2″ N	13° 23′ 56″ E	3
10	Baden-Wiirttemberg	Stuggart	Capital	48° 46′ 43″ N	9° 10′ 46″E	8
11	Bavaria	Munchen	Capital	48° 31′ 52″ N	11°57′ 50″E	5
12	Thuringia	Erfurt	Capital	50° 59′ 0″ N	11° 2′ 0″ E	1
13	Rhineland-Palatinate	Mainz	Capital	50° 0′ 0″ N	S° 16′ 16″ E	1

Table 3.2 The data table of the Escherichia coli epidemic in May 2011

Town	Longitude	Latitute
Hamburg	10.00139	53.56528
Bremen	8.8075	53.07583
Schleswig–Holstein	10.14056	54.32528
Mecklenburg-Vorpommern	11.41667	53.63333
Hesee-Frankfurt	8.682222	50.11028
Sarrland	7	49.23333
Lower_Saxony	9.716667	52.36667
North_Rhine-Westphalia	6.783333	51.23333
Berlin	13.39889	52.50056
Baden0Wuttemberg	9.179444	48.77861
Bavaria	11.96411	48.53116
Thuringa	11.03333	50.98333
Rhineland-Palatinate	8.271111	50

TWC only considers the last two columns of this table; the city names for TWC are empty labels without semantics

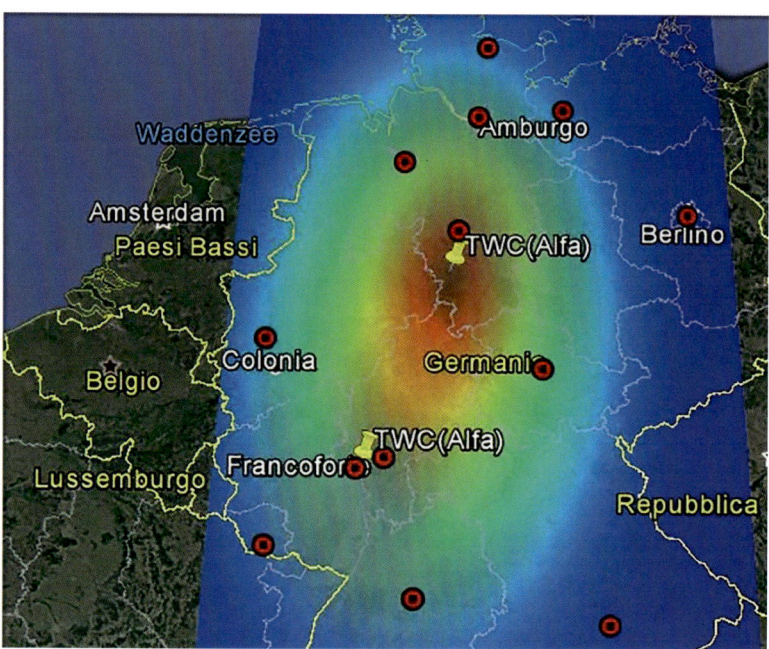

Fig. 3.5 The two yellow markers indicate the two locations from which. according to TWC. the epidemic is likely to have originated. The heat map (from blue to dark red) highlights the most relevant area of this spread

Figure 3.5 shows, with a heat map from blue to dark red and two markers (in yellow), the locations from which the epidemic is most likely to have originated according to TWC. In short, the city of Hanover (Lower Saxony) is the most probable followed by Frankfurt.

The challenge lies in verifying these hypotheses that go against the more established theories at that time (May 2011). Simply waiting for a few months may be sufficient.

The document "EHchEC O104:H4 Outbreak" by the Robert Koch Institute (Germany 2011): "Analysis of a **satellite outbreak in two canteens of a Frankfurt-based company** stated the following. Between 9 and 17 May 2011. a total of 60 employees at two locations of a Frankfurt-based company contracted bloody diarrhea, nine of which were laboratory-confirmed (to be EHEC infected), 18 of the 60 employees developed HUS. On 19 May 2011, the health department of the city of Frankfurt am Main was informed of the events by the personnel office of the company and initiated an investigation of the outbreak. The Health Protection Authority City and the Veterinary Service of Frankfurt am Main together with the Robert Koch Institute and the operators of the canteens were able to acquire a list of the purchases made in the canteen in the weeks from 2 to 16 May for the case of the persons ($n = 23$) and for randomly selected healthy persons ($n = 30$) with the help of electronic billing documentation. A logistic regression analysis was performed. The risk of contracting bloody diarrhea for employees who had bought and consumed a salad in the canteen in the above-mentioned period was six times higher compared to employees who had not bought any salad. A total of 20 of the 23 cases (87%) could be explained by the salad purchase. The consumption of other foods from the canteen was not significantly associated with the disease. With this study a list of salad bar items was made available in the canteen in order that the most likely vehicle could be identified. In the period in which the groups dined in the restaurant, the restaurant used only one mixture of sprouts containing fenugreek sprouts, alfalfa sprouts, adzuki bean sprouts, and lentil sprouts. **The supplier of sprouts for the restaurant received the sprouts from Company A in Lower Saxony responsible for the outbreak."**

3.2.2 Europe 2014: Terrorist Attacks

There were 92 terrorist attacks in Europe in 2014 (source: Global Terrorism Database. Start Project—University of Maryland). A data set containing only the coordinates of these 92 points may be considered small and apparently not very informative. However, by applying the TWC algorithm to this small but dense data set, it is possible to make hypotheses about the location of the logistical base of these attacks, the places of future attacks, and the geographic area from which most of these attacks were planned. Figure 3.6 shows the distribution of the 92 attacks in Europe in 2014. It can be noted that Germany and France were still rare targets while Belgium had no terrorist attacks during that year.

Fig. 3.6 The locations of the 92 terrorist attacks in Europe in 2014 (markers in blue)

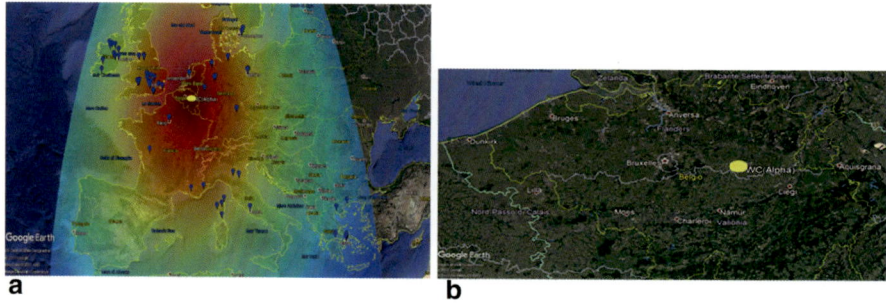

Fig. 3.7 a, b On the left. Heat map (probability) of the location of the logistical base of the attacks and the most probable area of its position (in yellow); on the right. zoom on the area of interest

The TWC algorithm was able to locate the possible logistical base(s) of the attacks in Belgium, east of Brussels (see Fig. 3.7a, b). Throughout Belgium there were no terrorist attacks throughout 2014 and in the same year attacks in France and Germany were rare (Fig. 3.6). However, the TWC heat map predicts an expansion of attacks in Germany and eastern France (Fig. 3.7a). In fact in mid-2016 a series of terrorist attacks began in Germany (Berlin and Munich). while in France the most brutal attacks began in 2015 (November 13, 2015 at the Bataclan and in six different areas of Paris, and on March 22. 2015, at the airport and metro in Brussels).

The media (Sky Med) report the news as follows:

> Six guilty of murder over 2016 Brussels airport and train attack that killed 32 people. Just four months after the Paris attacks, dozens more innocent people were killed in twin bombings in the Belgian capital. Six men have been found guilty of murder over the 2016 Brussels terror attacks that left 32 people dead. Bombs exploded in the airport and on a metro train passing through the city's European quarter in attacks claimed by Islamic State. Fifteen men and 17 women were killed with more than 300 people injured. The attacks on 22 March 2016 were the deadliest in Belgium since the end of the Second World War.

From what was reconstructed by the judges it was clear that a terrorist cell had been organizing in Brussels for some time and TWC detected this logistic point one year before the attacks. However, it is necessary to understand whether it is possible to derive additional information from such a reduced data set. For example, can TWC, using only the coordinates of the 92 locations where a terrorist attack occurred, make a credible hypothesis about the area from which these attacks were planned, perhaps not just the simple logistical base that acted as the focal point? Figure 3.8 shows the heat map that TWC processed to identify the area that best explains the geographical distribution of the 92 attacks. In this case an area between Birmingham and Manchester is indicated.

Below are some news from the main media between 2014 and 2015

a. Reuter **January 12, 2015** 3:51 PM UPDATED 9 YEARS AGO Birmingham: Britain's new Mecca?
b. Birmingham City University, Prevent Duty: The prevent duty which came into force on **September 18, 2015** is the statutory obligation on certain bodies.

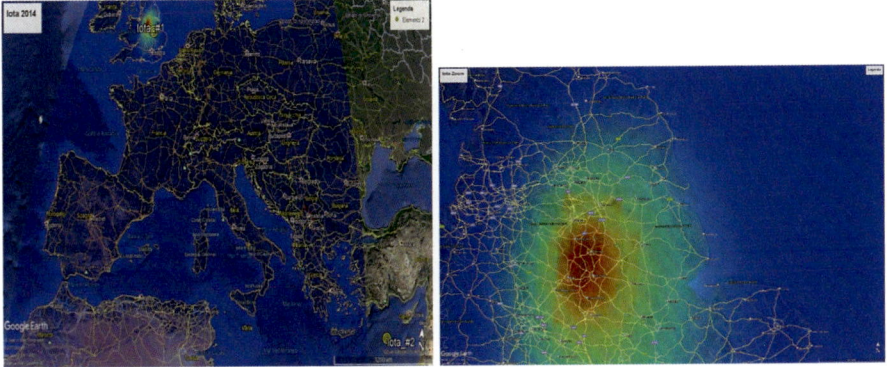

Fig. 3.8 On the left. TWC (Iota) displays the heat map that identifies the area that best explains the geographical distribution of the 92 attacks; on the right. a detailed view of the area between Birmingham and Manchester

    including universities. to "have due regard to the need to prevent people from being drawn into terrorism". The University takes the risk of radicalisation and threat of violent extremism and terrorism very seriously, yet at the same time strives to ensure that the institution strikes an appropriate balance between the duty and other statutory duties such as the freedom of speech (Education Act [No 2] 1986). Academic freedom (Education Reform Act 1988) and the Equality Act 2010, are all central to the ethos of Birmingham City University as a vibrant teaching and learning institution.

c.   BBC News. **December 5 2014**: Two British men who went to Syria to join rebel fighters have been jailed for 12 years and eight months each. Yusuf Sarwar and Mohammed Nahin Ahmed. both aged 22 and from Birmingham were sentenced for engaging in conduct in preparation of terrorist acts. The judge imposed an extended sentence period of five years. Sentencing Ahmed and Sarwar earlier, Judge Michael Topolski described the two men as "deeply committed to violent extremism".

d.   The Guardian, **May 10, 2016**: A mock terrorist attack has been carried out at one of the UK's busiest shopping centres, in a marauding assault similar to the Paris and Brussels atrocities. More than 800 volunteers took part in the staged attack at the Trafford Centre in Manchester on Monday night. As part of the drill a fake suicide bomber detonated an explosive device in a packed food court at the shopping centre.

e.   BBC **4 March 2015**: Abid Naseer: Terrorist plotted a Manchester bombing.

    We have shown only two concise examples of what is possible to achieve with "small" data sets that may seem uninformative at first glance.

References

1. Buscema M (2006), Sistemi ACM e Imaging Diagnostico – Le immagini mediche come Matrici Attive di Connessioni, [ACM Systems and Diagnostic Imaging-Medical Images as Active Connection Matrix] Springer, Milan
2. Buscema M, Catzola L, Grossi E (2008) Images as active connection matrixes: the J-net system. ICMED J 2(1):27–53
3. Buscema M, Grossi E (2010) J-net system: a new paradigm for artificial neural networks applied to diagnostic imaging, Chap. 16, pp 431–456. In: Capecchi V, Buscema B, Contucci P, D'Amore B (eds) Applications of mathematics in models, artificial neural networks and arts. https://doi.org/10.1007/987-90-481-8581-8, Springer Scienze+Business Media
4. Buscema M, Passariello R, Grossi E, Massini G, Fraioli F, Serra G (2013) J-net: an adaptive system for computer-aided diagnosis in lung nodule characterization. Chapter 2, pp 25–61. In: Tastle WJ (ed) Data mining applications using artificial adaptive systems, Springer Science+Business Media New York 2013. https://doi.org/10.1007/978-1-4614-4223-3_1
5. Alicandro M, Dominici D, Buscema PM (2018) A new enhancement filtering approach for the automatic vector conversion of the UAV photogrammetry output. Springer Nature Switzerland AG 2018 Ioannides M et al (eds) EuroMed 2018, LNCS 11196, pp 1–10. https://doi.org/10.1007/978-3-030-01762-0_26
6. Dominici D, Zollini S, Alicandro M, Della Torre F, Buscema M, Baiocchi V (2020) High resolution satellite images for instantaneous shoreline extraction using new enhancement algorithms. Geosciences. https://doi.org/10.3390/geosciences9030123
7. Amato M, Buscema M, Massini G, et al (2021) Assessment of new coronary features on quantitative coronary angiographic images with innovative unsupervised artificial adaptive systems: a proof-of-concept study. Front Cardiovasc Med 8, Article 730626. www.frontiers in.org
8. Buscema M, Massini G, Sacco P (2017) The topological weighted centroid (TWC): a topological approach to the time-space structure of epidemic and pseudo-epidemic processes. Phys A 492(2018):582–627
9. Buscema M, Sacco PL, Massin G et al (2018) Unravelling the space grammar of terrorist attacks: a TWC approach. Technol Forecast Soc Chang 132(2018):230–254. https://doi.org/10.1016/j.techfore.2018.02.006
10. Buscema M, Della Torre F (2019) Novel applications of spatial mapping to chemical or biological outbreaks. In: Chemical health threats: assessing and alerting. Royal Society of Chemistry
11. Buscema M, Asadi-Zeydabadi M, Lodwick W et al (2020) Analysis of the Ebola outbreak in 2014 and 2018 in West Africa and Congo by using artificial adaptive systems. Appl Artif Intell. https://doi.org/10.1080/08839514.2020.1747770
12. Buscema PM, DellaTorre F, Breda M et al (2020) COVID-19 in Italy and estreme data mining. Phys A 557:124991
13. Asadi-Zeydabadi M, Buscema M, Lodwick W et al (2021) Analysis of COVID-19 pandemic in USA, using topological weighted centroid. Comput Biol Med 136:104670
14. Asadi-Zeydabadi M, Buscema M, Saffari-Parizi et al (2021) Analysis of COVID-19 outbreak in Iran by a suite of artificial adaptive system algorithms. Comput Sci Eng 2(2):165–179. https://doi.org/10.22124/cse.2021.19484.1012
15. Buscema PM, Asadi-Zeydabadi M, Massini G et al (2023) The topological weighted centroid: a new vision of geographic profiling. Stud Comput Intell 1095. https://doi.org/10.1007/978-3-031-28901-9_1

Chapter 4
Large and Powerful ANNs Versus Small, Numerous, and Diverse ANNs

Convolutional Neural Networks (CNNs) have proven to be one of the state-of-the-art systems in image understanding and other complex tasks where input patterns must undergo convolutions. CNNs have highlighted the "vertical" development of a classical ANN significantly increasing the number of processing layers between the input (its pattern) and the output (its correct classification). Its intermediate layers including convolutional, pooling, and dropout layers are inspired by how the visual cortex processes light signals. However, the technique of error backpropagation and the synaptic adaptability of inhibitory connections do not find a counterpart in the physiology of the human brain. Furthermore, current CNNs feature an increasingly higher number of layers exceeding what is found in the human cortex as if they were ANNs from the 1980s that have undergone intensive steroid treatment (see Fig. 4.1).

CNNs can be summarized in a few key concepts for the forward phase (convolutions, pooling, dropping, flat layers, and soft max) and a single concept for the learning phase (backpropagation). CNNs lack algorithm diversity, the equivalent of biodiversity in nature. CNNs are vertically deep architectures but horizontally shallow/narrow. In biology it's biodiversity that creates the richness that offers unexpected solutions in the face of sudden and/or new events. Algorithm diversity is the mathematical counterpart of biodiversity in the mathematical and artificial world. It's the mathematical diversity of algorithms used in the learning phase that allows the same problem (data set) to be observed and evaluated from different and often orthogonal perspectives. Algorithm diversity is achieved through distributed learning that employs different algorithms, mathematically and topologically, on the same data foundation. That is, not all algorithms are gradient descent-based with error backpropagation "seasoned" with different "sauces" (ADAM, and so on). An ANN based on vector quantization sees different features in the same data set than a backpropagation-based algorithm leading to different errors in blind testing. A similar principle applies to a decision tree (random forest for example) or a Bayesian network. In essence it's the diversity of learning laws and thus algorithms that creates a stereophonic array of viewpoints on the same "object" and diverse assessments

© The Author(s), under exclusive license to Springer Nature Switzerland AG 2025
P. M. Buscema et al., *AI: A Broad and a Different Perspective*,
SpringerBriefs in Computational Intelligence,
https://doi.org/10.1007/978-3-031-80600-1_4

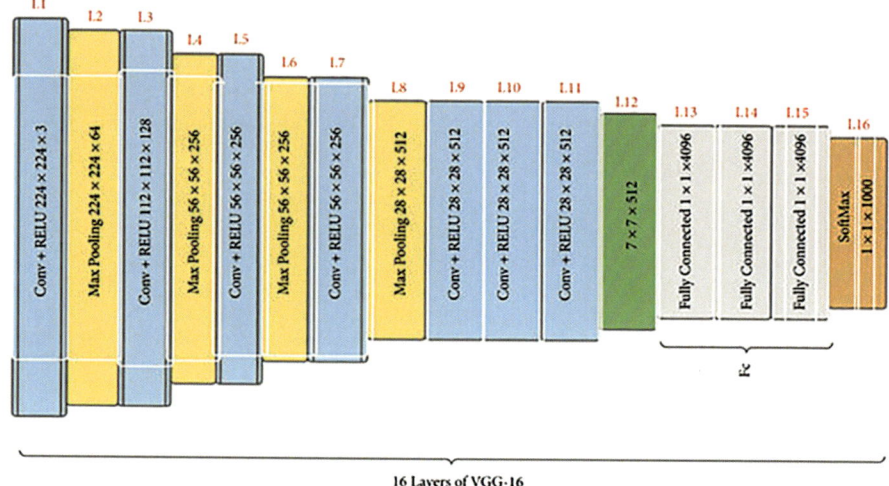

Fig. 4.1 The VGG-16, a typical CNN Architecture. Image taken from: Mohammed Rakeibu Hasan, Mohammed Ishraaf Fatemi, Mohammad Monirujjaman Khan, Comparative Analysis of Skin Cancer (Benign vs. Malignant) Detection Using Convolutional Neural Networks, Journal of Healthcare Engineering 2021(2):1–17

on "new objects", such that subsequent optimization can greatly enhance the final outcome. We propose therefore, to utilize this algorithm diversity to provide ANNs with not just vertical depth as found in standard CNNs, but also horizontal depth.

This strategic vision diverges from the traditional approach and is thus termed MESS (Multipolar Eco System Strategy). MESS analyzes any data set through a wide variety of adaptive algorithms, including CNNs. Once all the diverse and distributed algorithms have learned the same data set, each of them will express its evaluation for each new pattern used in the test set. Finally, an unsupervised Meta ANN (known as Meta Net), solely analyzing the confusion matrix that each algorithm generated during the training or validation phase, will create its specific weights through which it will definitively classify each new pattern in the test set. Meta Net trains and operates without any knowledge of the architecture of the algorithms constituting its "parliament". It is entirely unsupervised and thus remains unaware of the correct classification of each pattern in the test set (see Ref. for Meta Net: Buscema [1]; Buscema et al. [2, 3]).

Meta Net represents the final decision of a parliament of highly diverse set of algorithms each having demonstrated expertise in a distinct manner from the others. Meta Net and its algorithmic parliament provide the entire analytical system with significant horizontal depth and width due to their varied composition. In the realm of AI Meta Net and its algorithms present the "republican" solution to problem-solving (see Fig. 4.2) in contrast to the "monarchic" (in certain cases "oligarchic") solution embodied by a single large and powerful CNN.

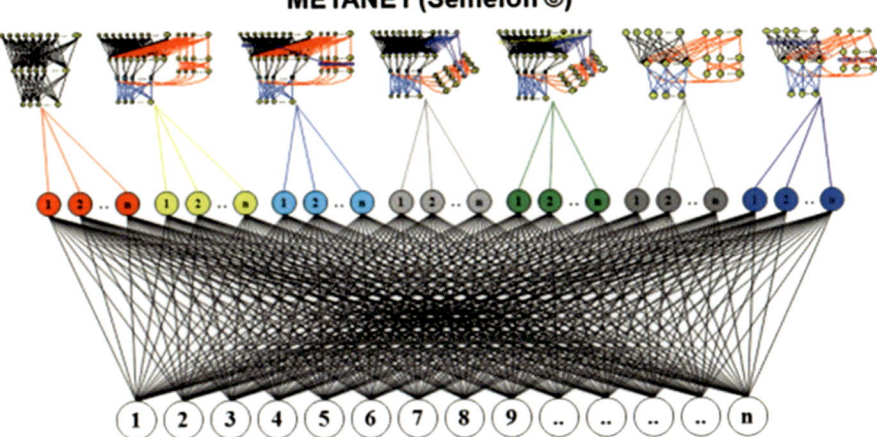

Fig. 4.2 An example of a "parliament" of different ANNs. Managed by Meta Net

We next experimentally demonstrate the superior performance that the MESS strategy can achieve compared to the classical approach. Applications involving image data sets are well-suited for this purpose. Thus let's exemplify the necessary steps to construct the MESS strategy (see also Fig. 4.3):

Preparation of a Parliament of ANNs.

a. Training-Validation-Testing of CNNs: Train and validate various CNN architectures either from scratch or pre-trained on an image data set and test all trained CNNs on a new image data set.
b. Extract features from CNNs: Utilize the values of the units in the flat layer of each CNN to rewrite the training, validation, and testing data sets.
c. Training-Validation-Testing of New ANNs and Algorithms: Use the new data sets composed of CNN features to train, validate, and test new ANNs and adaptive algorithms with deeply diverse architectures and learning laws.
d. Confusion Matrices: Extract confusion matrices from the training and validation results of all employed algorithms (CNNs. new ANNs. and algorithms).

Operation of Meta Net and its Parliament.

a. Generation of Meta Net Weights: Utilize the confusion matrices of all trained and validated algorithms to compose the complete weight matrices of the Meta Net ANN.
b. Meta Net Input: Organize all outputs of each testing pattern from each employed Algorithm (CNNs. New ANNs. and Algorithms) as the input vector for Meta Net.
c. Meta Net Output: Generate Meta Net's output for each testing pattern and compare the results with those already obtained in testing by all other algorithms.

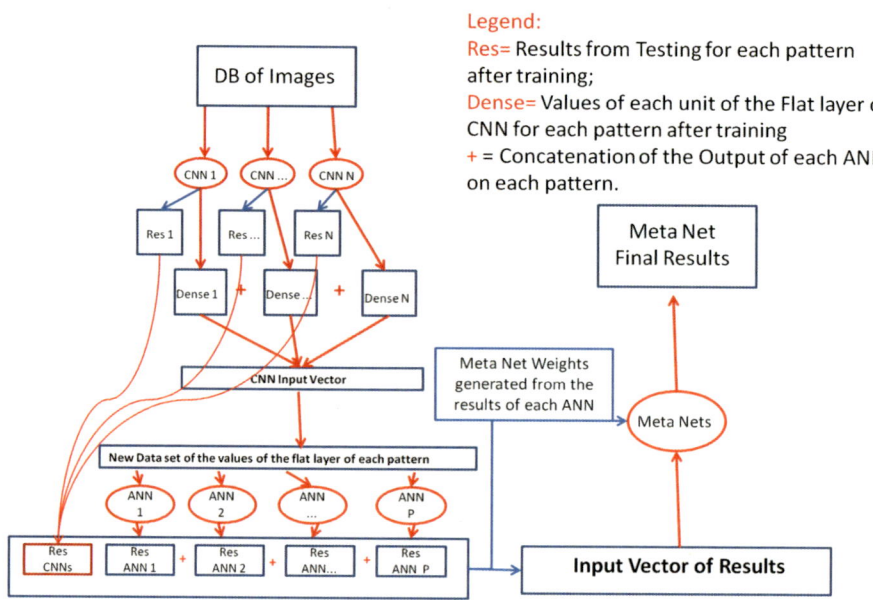

Fig. 4.3 Work flow according to MESS strategy for images analysis

4.1 MESS Versus Standard Approach: The MNIST Data Set

MESS is next benchmarked on the commonly used public data set MNIST and achieves astonishing results when compared to other published systems (Fig. 4.4).

The MNIST data set was divided, following the instruction of the authors, into training and testing set of 60 K and 10 K samples respectively. However, we randomly divided the training set into two splits of 50 K and 10 K samples. The former is used for training while the latter for validation. This is done to pick the best model up during learning [4].

Seven CNNs with different architectures and hyperparameters were trained (Fig. 4.5). Each of the seven CNNs was trained for 100 epochs using an exponential learning rate. On average training a CNN takes nearly an hour on a single NVIDIA GeForce GT 730M. The other ANNs trained on the flat units produced by each CNN required approximately 10 minutes of training.

The machine learning software TensorFlow was used for the training and evaluation of the CNNs. Most supervised networks and Meta-Nets were trained using the Semeion software suite: "Deep Supervised ANNs" v 35.3 (2023) and "Meta Net Multi Train" v 4.5 (2022). Other machine learning models were trained using the

MNIST

who is the best in MNIST ?

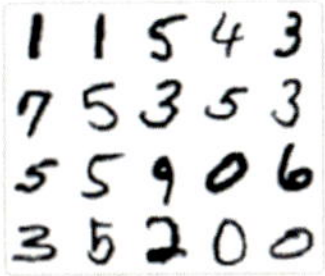

MNIST 50 results collected

Units: error %

Classify handwriten digits. **Some** additional results are available on the original dataset page.

Result	Method	Venue	Details
0.21%	Regularization of Neural Networks using DropConnect	ICML 2013	
0.23%	Multi-column Deep Neural Networks for Image Classification	CVPR 2012	
0.23%	APAC: Augmented PAttern Classification with Neural Networks	arXiv 2015	
0.24%	Batch-normalized Maxout Network in Network	arXiv 2015	Details
0.29%	Generalizing Pooling Functions in Convolutional Neural Networks: Mixed, Gated, and Tree	AISTATS 2016	Details
0.31%	Recurrent Convolutional Neural Network for Object Recognition	CVPR 2015	
0.31%	On the Importance of Normalisation Layers in Deep Learning with Piecewise Linear Activation Units	arXiv 2015	
0.32%	Fractional Max-Pooling	arXiv 2015	Details
0.33%	Competitive Multi-scale Convolution	arXiv 2015	
0.35%	Deep Big Simple Neural Nets Excel on Handwritten Digit Recognition	Neural Computation 2010	Details

Fig. 4.4 Best CNNs results on MNIST data set

CNN	0	1	2	3	4	5	6
Pre-processing	None	None	None	None	None	Crop(2,2)	Crop(2,2)
Conv-(4,4,20)	✓	✓	✓	✓	✓	✓	✓
Avg-pool-(2,2)	✓	✓	✓	✓	✓	✓	✓
Conv-(5,5,40)	✓	✓	✓	✓	✓	✓	✓
Avg-pool-(3,3)	✓	✓	✓	✓	✓	✓	✓
Conv-(2,2,32)	✗	✓	✓	✗	✗	✓	✓
Avg-pool-(2,2)	✗	✓	✓	✗	✗	✓	✓
Dense	150	150	150	150	128	✗	150
Data-Augment.	✗	✓	✓	✓	✓	✓	✓
Activation	relu	elu	relu	relu	relu	relu	relu

Fig. 4.5 Architecture of the seven CNNs trained on MNIST

software "WEKA." v 3.6 (2015).[1] In our experiments we use both well-known traditional methods as well as more recent and innovative classifiers from the literature (for references). We chose to use:

1. Convolutional Neural Networks (for short CNN) [5, 6];
2. Adaptive Vector Quantization (for short AVQ) [7–10];
3. Bayes Net [11];
4. Multilayer Back Propagation (for short BP) [12–14];
5. K Nearest Neighbourhood (for short KNN) [15, 16];
6. Logistic Regression [17, 18];
7. Naïve Bayes [19–21];
8. Random Forest [22, 23];
9. Sequential Minimal Optimization (for short SMO) [24–27];
10. Logit Boost [28].

In addiction we use the following ANNs designed at Semeion Research Center:

11. Bi-Modal ANN (for short Bm) [29];
12. Conic Net [30, 31];
13. Gauss Net (for short GNet) [31];
14. K Contractive Map (for short K-CM) [32–34];
15. New Recirculation Neural Network (for short NRC) [35];
16. Sine Net (for short Sn) [36–38];
17. Guacamole (for short Pop NRC) [39].

Each of these 15 algorithms was applied with various topologies, different hyper-parameters, and starting each time from the rewriting of the entire data set through the flat units of the different pre-trained CNNs.

Thus, the parliament of artificial systems consulted by Meta Net for making final decisions was composed of approximately 57 traditional and new ANNs and algorithms.

Meta Net [7], in summary, is a family of unsupervised meta classifiers methods combining the hypothesis coming from many different base classifiers to produce a powerful prediction rule. Each member of this family can be understood as a single layer perceptron whose weights are established without gradient based methodologies. Their input layer consists in the concatenation of the class probabilities produced by each of the classifiers in the ensemble. The output nodes represent instead the final unnormalized categorical distribution over classes used for prediction. Therefore, using K classifiers for a N-classes problem, we have KxN input nodes and N outputs nodes. The resulting weight matrix W of dimensions Nx(KxN) is generated by a fuzzy function of the sensitivity and precision of each of the base classifiers as estimated by their confusion matrices on the training set and of validation set.

Figure 4.6 illustrates how the Meta Net's weight matrix is generally computed from the confusion matrix of each classifier within its parliamentary members.

---

[1] Paolo Massimo Buscema, Francesca Della Torre, Antonio Loquercio.

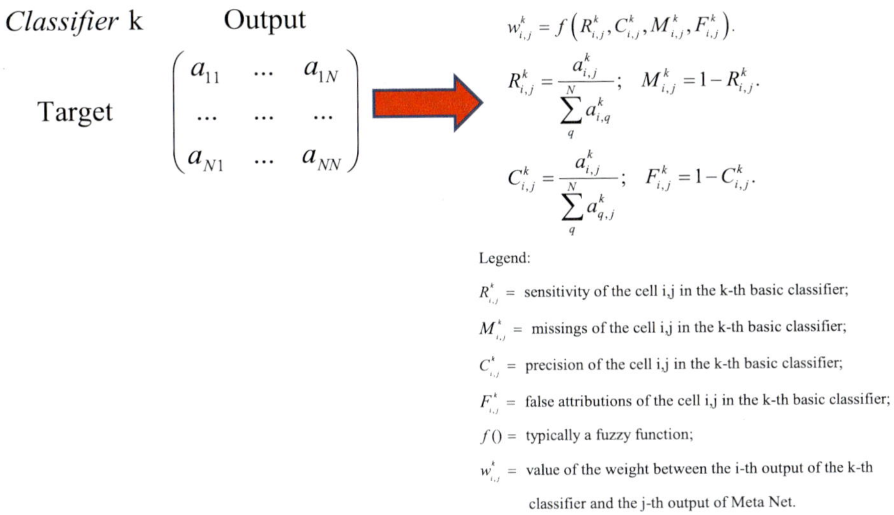

Classifier k   Output $w_{i,j}^k = f\left(R_{i,j}^k, C_{i,j}^k, M_{i,j}^k, F_{i,j}^k\right).$

Target
$$\begin{pmatrix} a_{11} & \cdots & a_{1N} \\ \cdots & \cdots & \cdots \\ a_{N1} & \cdots & a_{NN} \end{pmatrix}$$

$$R_{i,j}^k = \frac{a_{i,j}^k}{\sum_{q}^{N} a_{i,q}^k}; \quad M_{i,j}^k = 1 - R_{i,j}^k.$$

$$C_{i,j}^k = \frac{a_{i,j}^k}{\sum_{q}^{N} a_{q,j}^k}; \quad F_{i,j}^k = 1 - C_{i,j}^k.$$

Legend:

$R_{i,j}^k$ = sensitivity of the cell i,j in the k-th basic classifier;

$M_{i,j}^k$ = missings of the cell i,j in the k-th basic classifier;

$C_{i,j}^k$ = precision of the cell i,j in the k-th basic classifier;

$F_{i,j}^k$ = false attributions of the cell i,j in the k-th basic classifier;

$f()$ = typically a fuzzy function;

$w_{i,j}^k$ = value of the weight between the i-th output of the k-th

classifier and the j-th output of Meta Net.

Fig. 4.6 Procedure for each Meta Net to calculate the weights matrix involves deriving it from the confusion matrix of each classifier within the Meta Net parliament

This experimentation consisted of 6 types of different algorithms (function) implementing Meta Net that were applied each time with different function.

Meta Bayes developed in 2007 (see Ref. [5]) and it is shown in Eq. (4.1).It was inspired to the work of previous paper by McClelland and Hinton in 1986 (see Eqs. (4.2)):

$$w_{i,j}^k = f\left(R_{i,j}^k, C_{i,j}^k, M_{i,j}^k, F_{i,j}^k\right) = -\ln\left(\frac{M_{i,j}^k \cdot F_{i,j}^k}{R_{i,j}^k \cdot C_{i,j}^k}\right) = -\ln\left(\frac{(1 - R_{i,j}^k) \cdot (1 - C_{i,j}^k)}{R_{i,j}^k \cdot C_{i,j}^k}\right).$$
(4.1)

$$w_{i,j} = -\ln\left(\frac{p(x_i = 0 \,\&\, x_j = 1) \cdot p(x_i = 1 \,\&\, x_j = 0)}{p(x_i = 1 \,\&\, x_j = 1) \cdot p(x_i = 0 \,\&\, x_j = 0)}\right).$$
(4.2)

Meta Sum is the second type of Meta Net developed in 2007 (see Ref. [7]). Equation (4.3) shows the Meta Sum and Eqs. (4.4) its formulation in term of fuzzy set theory.

$$w_{i,j}^k = f\left(R_{i,j}^k, C_{i,j}^k, M_{i,j}^k, F_{i,j}^k\right) = \frac{1}{2} \cdot \left[\left(R_{i,j}^k + C_{i,j}^k\right) - \left((1 - R_{i,j}^k) \cdot (1 - C_{i,j}^k)\right)\right].$$
(4.3)

if we assume:

$$\tilde{A} = R_{i,j}^k \text{ and } \tilde{B} = C_{j,i}^k$$

Then Eq. (4.3) can be re written in fuzzy set terms as follow:

$$w_{i,j}^k = \frac{\mu_{\tilde{A}}(x^k) + \mu_{\tilde{B}}(x^k) - \left(1 - \mu_{\tilde{A}}(x^k)\right) \cdot \left(1 - \mu_{\tilde{B}}(x^k)\right)}{2}. \qquad (4.4)$$

Meta Fuzzy (see Eq. (4.5 and 4.6), **Meta Exp** [see Eq. (4.7) and **Meta Einstein** (see Eq. (4.8 and 4.9)] were developed few years later (see 3.Buscema et al. [2]). The Meta Einstein equation was inspired by fuzzy operator in fuzzy theory of sets (see Kahraman et al. [40]).

Meta Fuzzy

$$w_{i,j}^k = \max\left\{\min\left(R_{i,j}^k, F_{i,j}^k\right), \min\left(C_{i,j}^k, M_{i,j}^k\right)\right\} = \min\left\{R_{i,j}^k, C_{i,j}^k\right\}; \qquad (4.5)$$

In fuzzy terms:

$$\begin{aligned}
w_{i,j}^k &= \max\left\{\min\left(\mu_{\tilde{A}}(x^k), 1 - \mu_{\tilde{B}}(x^k)\right), \min\left(\mu_{\tilde{B}}(x^k), 1 - \mu_{\tilde{A}}(x^k)\right)\right\} \\
&= \min\left\{\mu_{\tilde{A}}(x^k), \mu_{\tilde{B}}(x^k)\right\}.
\end{aligned} \qquad (4.6)$$

Meta Exp

$$w_{i,j}^k = \frac{e^{\left(R_{i,j}^k + C_{i,j}^k\right)}}{e^{\left(M_{i,j}^k \cdot F_{i,j}^k\right)}}. \qquad (4.7)$$

Meta Einstein

$$w_{i,j}^k = \frac{R_{i,j}^k + C_{i,j}^k}{1 + M_{i,j}^k \cdot F_{i,j}^k}; \qquad (4.8)$$

In fuzzy terms:

$$w_{i,j}^k = \frac{\mu_{\tilde{A}}(x^k) + \mu_{\tilde{B}}(x^k)}{1 + \left(1 - \mu_{\tilde{A}}(x^k)\right) \cdot \left(1 - \mu_{\tilde{B}}(x^k)\right)}. \qquad (4.9)$$

The last member of the Meta Net family, **Meta Consensus**, was developed in 2010, presented and awarded in NAFIPS 2010 (Toronto) and inspired to Consensus Theory (see Wierman and Tastle [41]), and published in completed form in 2013 [3]. Equations (4.10–4.17) show how we have adapted the original equation of Consensus Theory.

Meta Consensus

$$r_{i,j}^k = \frac{a_{i,j}^k}{R_i^k} \cdot \log_2\left(R_i^k - \frac{\left|a_{i,j}^k - R_i^k\right|}{2 \cdot (N-1)}\right); \text{ where } R_i^k = \sum_j^N a_{i,j}^k. \tag{4.10}$$

$$c_{i,j}^k = \frac{a_{i,j}^k}{C_j^k} \cdot \log_2\left(C_j^k - \frac{\left|a_{i,j}^k - C_j^k\right|}{2 \cdot (N-1)}\right); \text{ where } C_j^k = \sum_i^N a_{i,j}^k. \tag{4.11}$$

$$m_{i,j}^k = \frac{R_i^k - a_{i,j}^k}{R_i^k} \cdot \log_2\left(R_i^k - \frac{a_{i,j}^k}{2 \cdot (N-1)}\right); \tag{4.12}$$

$$f_{i,j}^k = \frac{C_j^k - a_{i,j}^k}{C_j^k} \cdot \log_2\left(C_j^k - \frac{a_{i,j}^k}{2 \cdot (N-1)}\right); \tag{4.13}$$

$$\left\{\begin{array}{l} Agr(\mathbf{X},\tau) = 1 + \sum_{i=1}^n p_i \log_2\left(1 - \frac{|X_i - \tau|}{2 \cdot d_X}\right) = r_{i,j}^k + c_{i,j}^k; \\[2mm] Dagr(\mathbf{X},\tau) = -\sum_{i=1}^n p_i \log_2\left(1 - \frac{|X_i - \tau|}{2 \cdot d_X}\right) = m_{i,j}^k + f_{i,j}^k; \end{array}\right\} \tag{4.14}$$

$$y = Cns(\mathbf{X}) - Dnt(\mathbf{X}) = \left(r_{i,j}^k + c_{i,j}^k\right) - \left(m_{i,j}^k + f_{i,j}^k\right). \tag{4.15}$$

$$y^* = \left(\ln(r_{i,j}^k) + \ln(c_{i,j}^k)\right) - \left(\ln(m_{i,j}^k) + \ln(f_{i,j}^k)\right) = -\ln\left(\frac{m_{i,j}^k \cdot f_{i,j}^k}{r_{i,j}^k \cdot c_{i,j}^k}\right). \tag{4.16}$$

$$w_{i,j}^k = -\ln\left(\frac{m_{i,j}^k \cdot f_{i,j}^k}{r_{i,j}^k \cdot c_{i,j}^k}\right). \tag{4.17}$$

The formulation of diverse equations within the framework of Meta Net constitutes not merely a superficial endeavor, but rather assumes a foundational significance aimed at endowing Meta Net optimization processes with internal heterogeneity akin to "bio diversity." Such diversification within Meta Net is pivotal, as it enables the system to manifest varied estimations for novel patterns encountered, thereby enriching its adaptive capacity and enhancing its efficacy across a spectrum of contexts.

The results on the 10,000 prediction patterns are shown in Table 4.1.

One of the most important conclusions is the absence of a linear correlation between the accuracy of an algorithm and its probability of being selected for an optimal ensemble by Meta Net. In Fig. 4.7 in fact, the correlation between the percentage of times a classifier is selected by Meta Net to build an ensemble is presented together with its accuracy. Each blue dot represents a base classifier while the green curve is a 5th degree polynomial fitting to the points. Its behavior indicates

Table 4.1 Results on the MNIST data set of Meta Net ANNs and classifiers within its parliamentary assembly (Bold = Meta Nets; Italic = Initial CNNs; Normal = Other ANNs and Learning Machines)

Rank	ANN name	W.Mean	Error	Rank	ANN name	W.Mean	Error
1	**Meta-Einstein**	**0.9983**	**17**	34	(CNN4)FF_Bp24	0.9952	48
2	**Meta-Sum**	**0.9983**	**17**	35	(CNN4)FF_Bp64	0.9952	48
3	**Meta-Bayes**	**0.9982**	**18**	36	(CNN4)kNN3	0.9952	48
4	**Meta-Consensus**	**0.9982**	**18**	37	(CNN4)SMO	0.9952	48
5	**Meta-Expn**	**0.998**	**20**	38	(CNN4)Conic64	0.9951	49
6	**Meta-Fuzzy**	**0.998**	**20**	39	(CNN0)AVQ-Base_G48	0.995	50
7	(CNN2)FF_Bp(64)	0.9969	31	40	(CNN0)RotationForest	0.995	50
8	(CNN2)RandomForest	0.9969	31	41	(CNN4)FF_Bp86	0.995	50
9	(CNN2)FF_Bm(48)	0.9968	32	42	(CNN4)FF_Bp48	0.9949	51
10	*CNN2*	*0.9968*	*32*	43	(CNN4)RandomForest	0.9949	51
11	(CNN2)SMO	0.9967	33	44	(CNN4)NaiveBayes	0.9948	52
12	CNN6	0.9967	33	45	(CNN0)FF_Bp48	0.9946	54
13	(CNN2)Conic(64)	0.9965	35	46	(CNN3)FF_Bp64	0.9946	54
14	*CNN1*	*0.9963*	*37*	47	(CNN2)LogitBoost	0.9945	55
15	*CNN5*	*0.9962*	*38*	48	(CNN3)Conic48	0.9944	56
16	(CNN0)GNet_Ad120	0.9961	39	49	(CNN3)SMO	0.9944	56
17	(CNN2)K-CM5	0.9961	39	50	(CNN3)FF_Sn64	0.9943	57
18	(CNN4)FF_Bm48	0.9961	39	51	(CNN3)K-CM(3)	0.9943	57
19	(CNN0)FF_Bm88	0.996	40	52	(CNN0)LogitBoost	0.9942	58
20	(CNN4)NRC(64 × 48)	0.996	40	53	(CNN2)BayesNet	0.9941	59
21	MajorVote	0.996	40	54	(CNN3)RandomForest	0.994	60
22	*CNN0*	*0.9960*	*40*	55	(CNN4)BayesNet	0.994	60
23	(CNN0)Conic88	0.9959	41	56	(CNN3)FF_Bm64	0.9937	63
24	(CNN2)kNN5	0.9959	41	57	(CNN3)NaiveBayes	0.9937	63
25	*CNN4*	*0.9959*	*41*	58	*CNN3*	*0.9936*	*64*
26	(CNN0)RandomForest	0.9958	42	59	(CNN3)BayesNet	0.9934	66
27	(CNN0)SMO	0.9958	42	60	(CNN2)NaiveBayes	0.9933	67
28	(CNN0)FF_Sn120	0.9957	43	61	(CNN3)Pop_NRC	0.993	70
29	(CNN0)Conic64	0.9956	44	62	(CNN0)BayesNet	0.9928	72
30	(CNN0)kNN3	0.9955	45	63	(CNN0)NaiveBayes	0.9924	76
31	(CNN4))FF_Sn64	0.9955	45	64	(CNN4)Logistic	0.9901	99

(continued)

Table 4.1 (continued)

Rank	ANN name	W.Mean	Error	Rank	ANN name	W.Mean	Error
32	(CNN0)FF_Bp64	0.9953	47	65	(CNN3)Logistic	0.989	110
33	(CNN4)K-CM3	0.9953	47				

Each ANN, except the base CNNs, is followed by a number specifying the number of hidden units and layers used for training ((CNN4)FF_Sn64 = Sine Net with 1 layer of 64 hidden units trained on the flat layer of CNN 0) while the other learning machines are followed by the parameters used ((CNN3 K-CM(3) = K Contractive Map with 3 nearest neighbors trained on the flat layer of CNN3)

that the average accuracy gives little contribution to the creation of a performing classification committee. As can be observed in Fig. 4.7. Classification hypothesis with average accuracy has little influence on the final prediction rule. Instead, classifiers with an accuracy of 1 standard deviation (0.15%) from the average are fundamental for the entire artificial ecology. This is the experimental demonstration that weak classifiers can be crucial in Meta Nets' parliament. Weak classifiers very often estimate correctly some few patterns that the best classifiers misclassify. Consequently, the best classifiers and the worst sometimes have to belong to the same "elite" class.

The results of the Meta Net ANNs are explained by how these specific Meta Nets manage the diversity of the classifiers they utilize. In fact, it is plausible that each classifier makes different errors the more diverse its topology and the underlying mathematics of its learning are. Evidence of this can be seen in the graph in Fig. 4.8. This graph has been constructed based on the similarity of errors made by each classifier including Meta Nets, on the prediction set. The more two classifiers make

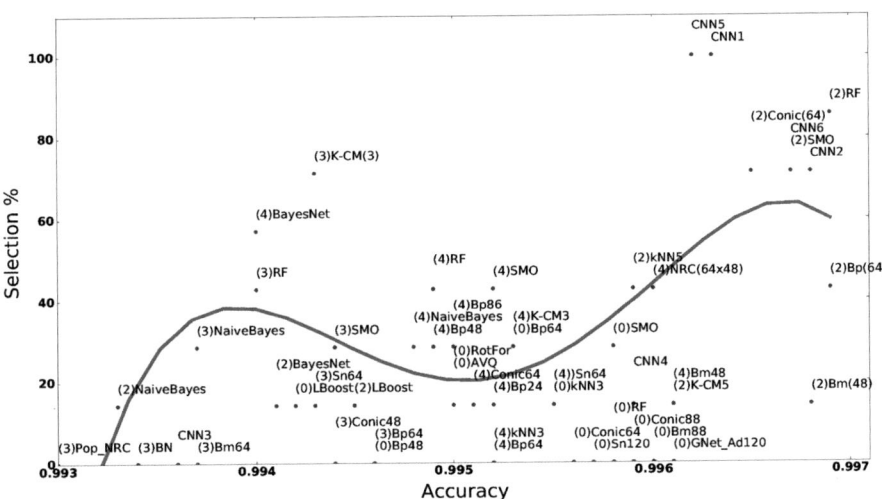

Fig. 4.7 The x-axis is the accuracy in prediction of each classifier member of the Meta Nets parliament and y-axis is the number of times each classifier has been selected by all 6 Meta Nets

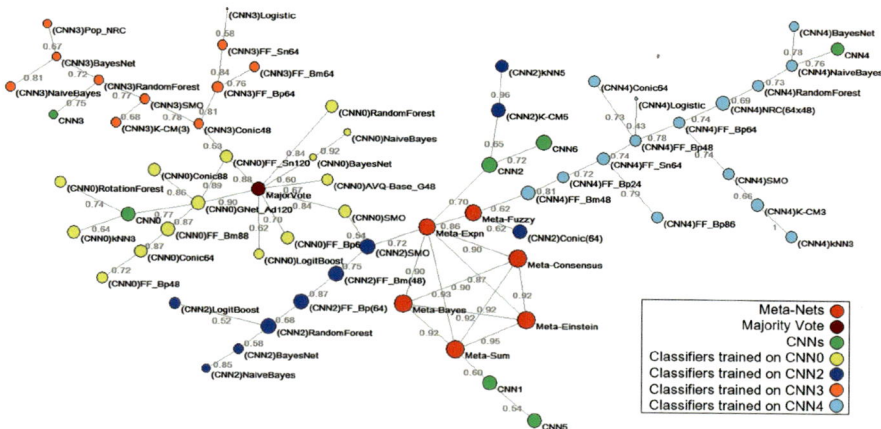

Fig. 4.8 In this graph each node represents a classification algorithm while the edge weights indicate the relationship's strength between nodes. To allow a better understanding, the size of the vertices has been drawn in proportion to the accuracy obtained by each algorithm

the same errors on the same patterns, the more similar their behavior is. Thus, the two nodes representing them are close and strongly connected.

It is evident from the graph how the family of Meta Net ANNs forms a self-contained and highly interconnected cluster while the other classifiers, even if close to one of the 6 Meta Nets, exhibit a much lower similarity value with these ANNs. Furthermore, it can be observed that the starting CNNs often generate much more efficient "offspring". This demonstrates that the learning of different algorithms on the rewriting that CNNs perform on the images (flat layer) is very effective. Additionally, it is noticeable that ANNs like K-CM and algorithms like Naive Bayes are distant in the graph from the Meta Nets even though these two "weak" algorithms in terms of predictive performance have been selected as fundamentals many times to enhance the results of Meta Net.

Figure 4.9 shows the comparison between the percentage of errors committed on the testing data set of the best Meta Net. a group of human experts and the best CNNs recorded up to 2015.

4.2   MESS Versus Standard Approach: Medical Applications

The MESS strategy has been compared with classical strategies for the classification of radiological images. In this case the procedure differed from the previous one. Pretrained CNNs (VGG16. RESNET18. INCEPTIONV3. ALEXNET. NOISYSTU-DENT) were chosen and a fine-tuning process was executed using the selected data

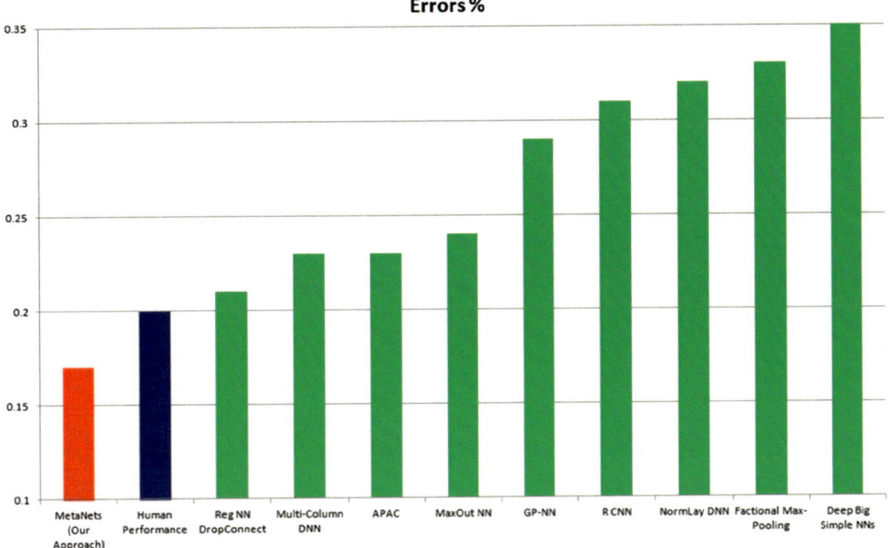

Fig. 4.9 Percent of errors committed on the MNIST data set by Meta Net (red), human expert panel (blue), and CNNs (green)

sets for the experiment. Subsequently, through an algorithm known as TWIST the most significant variables were extracted from the flat layer of each CNN. Traditional algorithms that are diverse in nature (kNN, Random Forest RF, Logistic Regression LR, Support Vector Classifier SVC, and Multi-Layer Perceptron MLP) were then trained on this new data set of extracted variables. The validation procedure employed was K-fold Cross Validation (with K = 5) for each experiment with a third test sample set aside for performance evaluation of the different models. By combining the results of Cross Validation from all CNNs and traditional algorithms, new weights for the Meta Net were generated. The Meta Net was then tested on an independent testing sample (see Fig. 4.10) for verification. The ensuing experiments were designed and conducted by researchers from the Semeion Research Center and the Bracco Group in 2022.[2]

---

[2] Semeion: Paolo Massimo Buscema, Giulia Massini; Bracco: Greta Clementi, Piero Bertani, Alessandro Maiocchi.

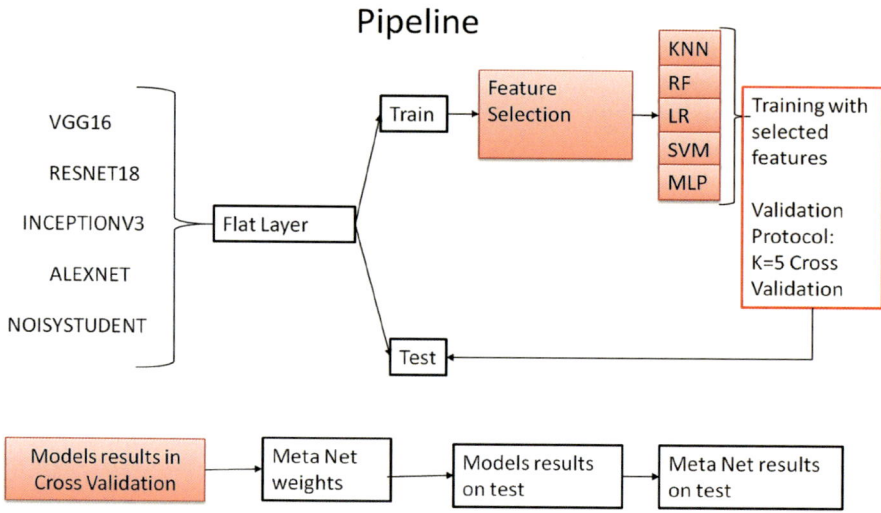

Fig. 4.10 Work flow of the experiments

4.2.1 Experiment 1: COVID-19 Radiography Database

This experiment involved a data set of 21,165 medical images categorized into 4 classes:

Class_1: 3.616 COVID-19 patterns.
Class_2: 6.012 Lung Opacity patterns.
Class_3: 10.192 Normal patterns.
Class_4: 1.345 Pneumonia patterns.
    [*Source*   https://www.kaggle.com/datasets/tawsifurrahman/covid19-radiog raphy-database].

Figure 4.11 displays the outcomes achieved on the independent testing set by CNNs, classical algorithms, and Meta Net (Meta Bayes).

The use of Meta Net enables a more accurate classification by 11 percentage points compared to the best algorithm (Meta Net: 157 errors. RESNET18 SVC: 610 errors).

4.2.2 Experiment 2: Chest X-Ray Images (Pneumonia)

This experiment involved a dataset of 5856 medical images. Categorized into 2 classes:

Class_1: 1.583 Normal Lungs.
Class_2: 4.273 Lungs with Pneumonia.

TEST	Output#1	Output#2	Output#3	Output#4	A.Mean	W.Mean	Error
model_ALEXNET_SVC.nnr	0.7469	0.7506	0.8965	0.7918	0.7965	0.8228	750
model_INCEPTIONV3_LogisticRegression.nnr	0.4869	0.7714	0.871	0.8364	0.7414	0.7749	953
model_INCEPTIONV3_MLPClassifier.nnr	0.668	0.7182	0.7959	0.8476	0.7574	0.7553	1036
model_INCEPTIONV3_SVC.nnr	0.6349	0.7697	0.9038	0.8662	0.7936	0.8174	773
model_NOISYSTUDENT_MLPClassifier.nnr	0.7842	0.7448	0.8224	0.8848	0.809	0.7978	856
model_NOISYSTUDENT_SVC.nnr	0.7815	0.7648	0.9274	0.8625	0.834	0.8521	626
model_RESNET18_KNeighborsClassifier.nnr	0.6584	0.7091	0.8871	0.8625	0.7793	0.7959	864
model_RESNET18_MLPClassifier.nnr	0.7856	0.7739	0.8405	0.8885	0.8221	0.8153	782
model_RESNET18_SVC.nnr	**0.769**	**0.7814**	**0.9254**	**0.8959**	**0.8429**	**0.8559**	**610**
model_VGG16_MLPClassifier.nnr	0.6777	0.6908	0.7993	0.803	0.7427	0.7479	1067
model_VGG16_RandomForestClassifier.nnr	0.5505	0.7057	0.8346	0.5874	0.6696	0.7338	1127
model_VGG16_SVC.nnr	0.7026	0.7423	0.9009	0.7844	0.7826	0.8146	785
MetaNet	**0.9322**	**0.921**	**0.9956**	**0.9851**	**0.9585**	**0.9629**	**157**

Fig. 4.11 Results on the COVID-19 Testing Set (Meta Net in red is the best result from one of the CNN members of the Meta Net parliament in bold)

[*Source* https://www.kaggle.com/datasets/paultimothymooney/chest-xray-pneumonia].

Figure 4.12 displays the outcomes achieved on the independent testing set by CNNs, classical algorithms, and Meta Net (Meta Bayes).

The use of Meta Net enables a more accurate classification by 4.5 percentage points compared to the best algorithm (Meta Net: 3 errors. NOISYSTUDENT SVC: 58 errors).

4.2.3 Experiment 3: Brain Tumor MRI Dataset

The data set comprises 7023 medical image categorized into 4 classes:

Class_1: 1.621 Glioma patterns.
Class_2: 1.645 Meningioma patterns.
Class_3: 2.000 No Tumor patterns.
Class_4: 1.757 Pituitary patterns.
[*Source* https://www.kaggle.com/datasets/masoudnickparvar/brain-tumor-mri-dataset].

TEST	Output#1	Output#2	A.Mean	W.Mean	Error
model_ALEXNET_KNeighborsClassifier.nnr	0.8864	0.9462	0.9163	0.9300	82
model_ALEXNET_LogisticRegression.nnr	0.8801	0.9591	0.9196	0.9377	73
model_ALEXNET_MLPClassifier.nnr	0.8770	0.9520	0.9145	0.9317	80
model_ALEXNET_RandomForestClassifier.nnr	0.8707	0.9485	0.9096	0.9275	85
model_ALEXNET_SVC.nnr	0.8959	0.9649	0.9304	0.9462	63
model_INCEPTIONV3_KNeighborsClassifier.nnr	0.8738	0.9251	0.8995	0.9113	104
model_INCEPTIONV3_LogisticRegression.nnr	0.8233	0.9497	0.8865	0.9155	99
model_INCEPTIONV3_MLPClassifier.nnr	0.8549	0.9532	0.9041	0.9266	86
model_INCEPTIONV3_RandomForestClassifier.nnr	0.7823	0.9404	0.8613	0.8976	120
model_INCEPTIONV3_SVC.nnr	0.8517	0.9614	0.9066	0.9317	80
model_NOISYSTUDENT_KNeighborsClassifier.nnr	0.8991	0.9497	0.9244	0.9360	75
model_NOISYSTUDENT_LogisticRegression.nnr	0.8991	0.9509	0.9250	0.9369	74
model_NOISYSTUDENT_MLPClassifier.nnr	0.9306	0.9567	0.9437	0.9497	59
model_NOISYSTUDENT_RandomForestClassifier.nnr	0.9211	0.9404	0.9307	0.9352	76
model_NOISYSTUDENT_SVC.nnr	**0.9306**	**0.9579**	**0.9442**	**0.9505**	**58**
model_RESNET18_KNeighborsClassifier.nnr	0.9117	0.9544	0.9330	0.9428	67
model_RESNET18_LogisticRegression.nnr	0.9117	0.9544	0.9330	0.9428	67
model_RESNET18_MLPClassifier.nnr	0.8864	0.9614	0.9239	0.9411	69
model_RESNET18_RandomForestClassifier.nnr	0.8991	0.9380	0.9185	0.9275	85
model_RESNET18_SVC.nnr	0.9180	0.9591	0.9385	0.9480	61
model_VGG16_KNeighborsClassifier.nnr	0.8486	0.9567	0.9027	0.9275	85
model_VGG16_LogisticRegression.nnr	0.8233	0.9754	0.8994	0.9343	77
model_VGG16_MLPClassifier.nnr	0.8170	0.9544	0.8857	0.9172	97
model_VGG16_RandomForestClassifier.nnr	0.7886	0.9579	0.8733	0.9121	103
model_VGG16_SVC.nnr	0.8391	0.9708	0.9049	0.9352	76
MetaNet	0.9905	1.0000	0.9953	0.9974	3

Fig. 4.12 Results on Pneumonia X-ray testing set (Meta Net in red, the best result from one of the CNN members of the Meta Net parliament in bold)

Figure 4.13 showcases the outcomes achieved on the independent testing set by CNNs, classical algorithms, and Meta Net (Meta Bayes).

Once again the use of Meta Net enables a more accurate classification by 4 percentage points compared to the best algorithm (Meta Net: 48 errors. NOISYSTUDENT kNN: 113 errors).

Summary: This chapter has provided experimental evidence of the effectiveness of diversity in both natural and artificial systems and highlighting the superiority of the MESS strategy over classic learning approaches. Additionally, we have showcased how perceived "weakness" can actually serve as a cornerstone for building robust solutions to a wide array of problems. Further comprehensive instances of Meta Nets applications and outcomes can be explored in dedicated research papers [REF].

TEST	Output#1	Output#2	Output#3	Output#4	A.Mean	W.Mean	Error
model_ALEXNET_KNeighborsClassifier.nnr	0.8765	0.7933	0.9350	0.9886	0.8984	0.9018	138
model_ALEXNET_LogisticRegression.nnr	0.7593	0.7052	0.9200	0.8665	0.8127	0.8192	254
model_ALEXNET_MLPClassifier.nnr	0.8426	0.8389	0.9675	0.9460	0.8988	0.9032	136
model_ALEXNET_RandomForestClassifier.nnr	0.7407	0.7842	0.9425	0.8750	0.8356	0.8420	222
model_ALEXNET_SVC.nnr	0.8025	0.8389	0.9650	0.9688	0.8938	0.8989	142
model_INCEPTIONV3_KNeighborsClassifier.nnr	0.8025	0.7842	0.9625	0.9659	0.8788	0.8847	162
model_INCEPTIONV3_LogisticRegression.nnr	0.8148	0.6170	0.9300	0.8892	0.8128	0.8199	253
model_INCEPTIONV3_MLPClassifier.nnr	0.8519	0.7781	0.9700	0.9545	0.8886	0.8940	149
model_INCEPTIONV3_RandomForestClassifier.nnr	0.7809	0.6596	0.9325	0.8807	0.8134	0.8206	252
model_INCEPTIONV3_SVC.nnr	0.7994	0.7720	0.9725	0.9574	0.8753	0.8819	166
model_NOISYSTUDENT_KNeighborsClassifier.nnr	**0.8704**	**0.8541**	**0.9625**	**0.9773**	**0.9161**	**0.9196**	**113**
model_NOISYSTUDENT_LogisticRegression.nnr	0.8086	0.6079	0.9325	0.9006	0.8124	0.8199	253
model_NOISYSTUDENT_MLPClassifier.nnr	0.8549	0.8815	0.9775	0.9716	0.9214	0.9253	105
model_NOISYSTUDENT_RandomForestClassifier.nnr	0.7562	0.7234	0.9575	0.9119	0.8373	0.8448	218
model_NOISYSTUDENT_SVC.nnr	0.8241	0.8602	0.9825	0.9744	0.9103	0.9153	119
model_RESNET18_KNeighborsClassifier.nnr	0.8889	0.8267	0.9600	0.9801	0.9139	0.9174	116
model_RESNET18_LogisticRegression.nnr	0.8210	0.6869	0.9325	0.9034	0.8360	0.8420	222
model_RESNET18_MLPClassifier.nnr	0.8735	0.8480	0.9725	0.9290	0.9057	0.9096	127
model_RESNET18_RandomForestClassifier.nnr	0.8179	0.6839	0.9575	0.8892	0.8371	0.8441	219
model_RESNET18_SVC.nnr	0.8395	0.8511	0.9700	0.9688	0.9073	0.9117	124
model_VGG16_KNeighborsClassifier.nnr	0.8148	0.7416	0.9550	0.9688	0.8701	0.8762	174
model_VGG16_LogisticRegression.nnr	0.7840	0.6170	0.9200	0.8636	0.7962	0.8036	276
model_VGG16_MLPClassifier.nnr	0.7994	0.7751	0.9575	0.9347	0.8667	0.8726	179
model_VGG16_RandomForestClassifier.nnr	0.7500	0.6140	0.9575	0.8523	0.7934	0.8028	277
model_VGG16_SVC.nnr	0.7994	0.7629	0.9825	0.9631	0.8770	0.8840	163
MetaNet	**0.9043**	**0.8967**	**0.9850**	**0.9886**	**0.9437**	**0.9466**	**75**

Fig. 4.13 Results on MRI tumor testing set (Meta Net in red, the best result from one of the CNN members of the Meta Net parliament in bold)

References

1. Buscema M (1998) Meta net: the theory of independent judges. SUM 33(2):439–461
2. Buscema M, Terzi S, Tastle W (2010) A new meta-classifier. In: Fuzzy information processing society (NAFIPS), 2010 annual meeting of the North American, 12–14 July 2010, Toronto, ON. IEEE, pp 1–7. https://doi.org/10.1109/NAFIPS.2010.5548298
3. Buscema M, Tastle WJ, Terzi S (2013) Meta net: a new meta-classifier family, Chapter 5. In: Tastle WJ (ed) Data mining applications using artificial adaptive systems. Springer Science+Business Media New York, pp 141–182
4. Loquercio, Della Torre F, Buscema M (2017) Computational eco-systems for handwritten digits recognition. arXiv:1703.01872v1 [stat.ML]

5. Goodfellow I, Bengio Y, Courville A (2016) Deep learning. MIT Press, Cambridge (MA)

6. Gulli A, Kapoor A, Pal S (2019) Deep learning with TensorFlow 2 and Keras, 2nd Ed, Packt, Pages, 646 pp. ISBN 9781838823412

7. Kohonen T (1990) Improved versions of learning vector quantization, 1st edn. International Joint Conference on Neural Networks, San Diego, pp 545–550

8. Kosko B (1992) Neural networks and fuzzy systems: a dynamical systems approach to machine intelligence. Prentice Hall, Englewood Cliffs, NJ

9. Kosko B (1992) Neural networks for signal processing. Prentice Hall, Englewood Cliffs, NJ

10. Buscema M, Catzola L (2010) AVQ1 basic and AVQ2 advanced, Semeion Report

11. Friedman N, Geiger D, Goldszmidt M (1997) Bayesian network classifiers. Mach Learn 29:131–163

12. Rumelhart DE, Hinton GE, Williams RJ (1986) Learning internal representations by error propagation. In: Rumelhart DE, McClelland JL (eds) Parallel distributed processing, vol 1. The MIT Press, Boston, pp 318–362

13. CHAUVIN (1995) In: Chauvin Y, Rumelhart DE (eds) Backpropagation: theory, architectures, and applications, lawrence erlbaum associates, Inc. Publishers 365 Brodway- Hillsdale, New Jersey

14. Buscema M (1998) Back propagation neural networks. Substance Use Misuse 33(2):233–270

15. Kowalski BR, Bender CF (1972) The K-nearest neighbor classification rule (pattern recognition) applied to nuclear magnetic resonance spectral interpretation. Anal Chem 44:1405–1411

16. Aha DW, Kibler D, Albert MK (1991) Instance-based learning algorithms. Mach Learn 6:37–66

17. Cessie S, van Houwelingen JC (1992) Ridge estimators in logistic regression. Appl Stat 41:191–201

18. Hosmer DW, Leneshow S (2000) Applied logistic regression, 2nd edn. Wiley, New York (NY, USA)

19. Zang H (2004) The optimality naive Bayes. Am Assoc Artif Intell. www.aaai.org

20. John GH, Langley P (1995) Estimating continuous distributions in Bayesian classifiers. In: Proceedings of the eleventh conference on uncertainty in artificial intelligence. Morgan Kaufmann Publishers, San Mateo

21. Rish I (2001) An empirical study of the naïve Bayes classifier. In: IBM research report, RC 22230 (W0111-014), New York

22. Breiman L (2001) Random forest. Mach Learn 45:5–32

23. Quinlan JR (1993) C4.5: programs for machine learning. Morgan Kaufman, San Mateo

24. Platt J (1998) Fast training of support vector machines using sequential minimal optimization. In: Schoelkopf B, Burges CJC, Smola AJ (eds) Advances in Kernel methods—support vector learning. MIT Press, Cambridge (MA, USA)

25. Keerthi SS, Shevade SK, Bhattacharyya C, Murthy KR-K (2001) Imrovements to Platt's SMO algorithm for SVM classifier design. Neural Comput 13:637–649

26. Keerthi SS, Gilbert EG (2002) Convergence of a generalized SMO algorithm for SVM classifier design. Mach Learn 46:351–360

27. Kecman V (2001) Learning and soft computing—support vector machines, neural networks, Fuzzy Logic systems. The MIT Press, Cambridge, MA

28. Otero J, Sánchez L (2006) Induction of descriptive fuzzy classifiers with the Logitboost algorithm. Soft Comput 10(9):825–835

29. Buscema M (2004) Bi Modal Networks, book chapter 9. In: Buscema M (ed), Reti Neurali Artificiali per l'orientamento professionale, Franco Angeli, Semeion Reading, Milano, pp 179–180

30. Buscema M, Massini G, Fabrizi M, Breda M, Della Torre F (2017) The ANNs approach to dem reconstruction. Comput Intell 1–35. https://doi.org/10.1111/coin.12151

31. Langer T, Favarato M, Buscema M (2020).Development of machine learning models to predict RT-PCR results for severe acute respiratory syndrome coronavirus 2 (SARSCoV-2) in patients with influenza-like symptoms using only basic clinical data. Scandinavian J Trauma, Resuscit Emerg Med 28:113. https://doi.org/10.1186/s13049-020-00808-8

32. Buscema M, Consonni V, Ballabio D, Mauri A, Massini G, Breda M, Todeschini R (2014) K-CM: a new artificial neural network. Application to supervised pattern recognition. Chemometr Intell Lab Syst 138:110–119
33. Alberoni M, Nemni R, Comi G, Buscema M, Furlan R, Grossi E (2015) A global immune deficit in Alzheimer's disease and mild cognitive impairment disclosed by a Novel data mining process. J Alzheimer's Disease 43:1199–1213. https://doi.org/10.3233/JAD-141116
34. Brancato A, Buscema PM, Massini G, Gresta S, Salerno G, Torre FD (2019) K-CM application for supervised pattern recognition at Mt. Etna: an innovative tool to forecast flank eruptive activity. Bull Volcanol 81:40. https://doi.org/10.1007/s00445-019-1299-4
35. Buscema M (1998) Recirculation neural networks. In: Buscema M (ed) Substance use and misuse 33(2):383—388. Special Issue on Artificial Neural Networks and Complex Social Systems
36. Buscema M, Terzi S, Breda M (2006) Using sinusoidal modulated weights improve feedforward neural network performances in classification and functional approximation problems. WSEAS Trans Inf Sci Appl 3(5):885–893
37. Buscema M (2010) Sine Net: an artificial neural network. Applicant Semeion Research Centre. Inventor M. Buscema. European Patent (Application n. 03425582.8 deposited 09–09–2003). USA Patent No US 7,788,196 B2 - Aug. 31, 2010. International Patent: Application PCT/EP2004/05189 deposited 08–28–2004
38. Buscema M, Terzi S, Breda M (2006) A feed forward sine based neural network for functional approximation of a waste incinerator emissions. In: Proceedings of the 8th WSEAS international conference on automatic control, modeling and simulation, Praga
39. Buscema M, Sacco PL (2013) GUACAMOLE: a new paradigm for unsupervised competitive learning, Chapter 7. In: Tastle WJ (ed) Data mining applications using artificial adaptive systems. Springer Science+Business Media New York, pp 211–230. https://doi.org/10.1007/978-1-4614-4223-3_1
40. Kahraman C, Öztayşi B, Onar SÇ (2016) A comprehensive literature review of 50 years of Fuzzy Set theory. Int J Comput Intell Syst 9(sup1):3–24. https://doi.org/10.1080/18756891.2016.1180817
41. Wierman MJ, Tastle WJ (2005) Consensus and dissention: theory and properties. NAFIPS 2005–2005 annual meeting of the North American Fuzzy information processing society, Detroit, MI, USA, pp 75–79. https://doi.org/10.1109/NAFIPS.2005.1548511

Chapter 5
Specialized Nodes Versus Conscious Nodes

Both supervised and unsupervised artificial neural networks each have nodes within the hidden layers that tend to specialize during the learning phase. This specialization involves encoding specific features of the observed patterns while disregarding others. This phenomenon is inherently linked to the technique of gradient descent and the chain rule used for backpropagating errors from the output layer. It's as if each node comprehends a fragment of an image. For example, a node might comprehend an the ears, another the trunk, or yet another the limbs of an elephant while no individual internal node possesses a holistic albeit fuzzy view of the entire elephant. This non-holistic nature of nodes in classical ANNs could be a constraint to overcome. A conscious node is the term we use for a node that has a holistic view rather a partial or local view.

This is illustrated in the following example. Figure 5.1 presents 13 patterns within a 7 × 7 square forming an oblique bar that moves from top to bottom and from right to left.

A classical autoencoder can learn this dataset in a few epochs. After training we explore the reconstruction capability of the autoencoder by activating specific pixels within the 7 × 7 square. Figure 5.1 demonstrates the response of the trained autoencoder when the top-left pixel is activated, pixel (1.1) is activated only once during training and yet the autoencoder can reconstruct the entire pattern going to the pattern to which it belongs (See Pat #7). Similar behavior occurs (Fig. 5.3) when two pixels are simultaneously activated in input: pixel (2.1) and pixel (3.2) (see Pat #6). In essence the trained autoencoder effectively fulfils its purpose by reconstructing original patterns from incomplete inputs.

This phenomenon arises due to the interconnecting weight matrices, which link each node (pixel) to every other, compelling each node to express or inhibit its activation in a specific manner based on how it was activated during training with complete patterns. Thus, each node in the trained autoencoder acquires partial knowledge of the entire data set. It lacks awareness of the training dataset's structural organization.

© The Author(s), under exclusive license to Springer Nature Switzerland AG 2025   49
P. M. Buscema et al., *AI: A Broad and a Different Perspective*,
SpringerBriefs in Computational Intelligence,
https://doi.org/10.1007/978-3-031-80600-1_5

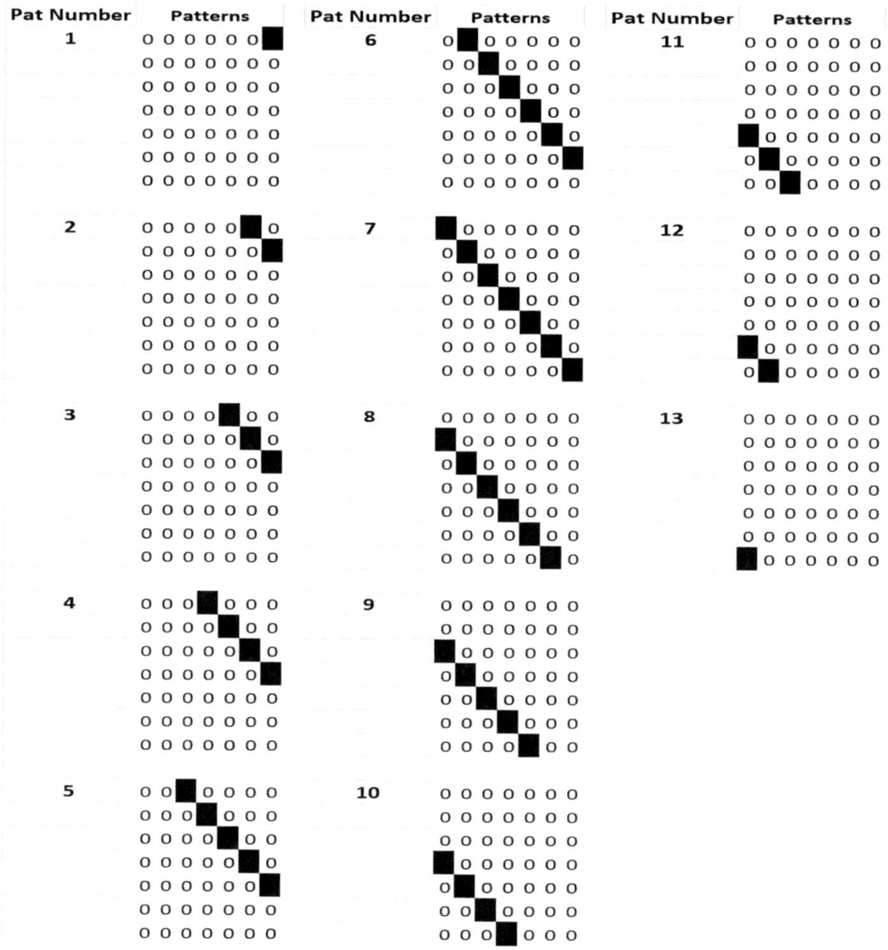

Pat Number	Patterns	Pat Number	Patterns	Pat Number	Patterns

1 — o o o o o o █ / o o o o o o o / o o o o o o o / o o o o o o o / o o o o o o o / o o o o o o o / o o o o o o o

6 — o █ o o o o o / o o █ o o o o / o o o █ o o o / o o o o █ o o / o o o o o █ o / o o o o o o █ / o o o o o o o

11 — o o o o o o o / o o o o o o o / o o o o o o o / o o o o o o o / █ o o o o o o / o █ o o o o o / o o █ o o o o

2 — o o o o o █ o / o o o o o o █ / o o o o o o o / o o o o o o o / o o o o o o o / o o o o o o o / o o o o o o o

7 — █ o o o o o o / o █ o o o o o / o o █ o o o o / o o o █ o o o / o o o o █ o o / o o o o o █ o / o o o o o o █

12 — o o o o o o o / o o o o o o o / o o o o o o o / o o o o o o o / o o o o o o o / █ o o o o o o / o █ o o o o o

3 — o o o o █ o o / o o o o o █ o / o o o o o o █ / o o o o o o o / o o o o o o o / o o o o o o o / o o o o o o o

8 — o o o o o o o / █ o o o o o o / o █ o o o o o / o o █ o o o o / o o o █ o o o / o o o o █ o o / o o o o o █ o

13 — o o o o o o o / o o o o o o o / o o o o o o o / o o o o o o o / o o o o o o o / o o o o o o o / █ o o o o o o

4 — o o o █ o o o / o o o o █ o o / o o o o o █ o / o o o o o o █ / o o o o o o o / o o o o o o o / o o o o o o o

9 — o o o o o o o / o o o o o o o / █ o o o o o o / o █ o o o o o / o o █ o o o o / o o o █ o o o / o o o o █ o o

5 — o o █ o o o o / o o o █ o o o / o o o o █ o o / o o o o o █ o / o o o o o o █ / o o o o o o o / o o o o o o o

10 — o o o o o o o / o o o o o o o / o o o o o o o / █ o o o o o o / o █ o o o o o / o o █ o o o o / o o o █ o o o

Fig. 5.1. 13 patterns representing an oblique bar traversing the 7×7 square at 13 instances

But through the trained/matured weights, it recognizes that when a node is activated alongside another node with which it was previously activated, it tends to reactivate.

Further experiments would make this evident. Activating Pixel (7.1) (see Pat #1) results in the trained autoencoder reproducing the same input as output since this pixel was not connected to any other pixel in the training dataset. Similar outcomes occur when Pixel (1.7) is activated or when both of these pixels are activated simultaneously (see Fig. 5.4a–c).

Consideration must also be given to how a classic autoencoder, once trained, would behave when confronted with an entirely new pattern. Figure 5.5 provides a meaningful response to this question. The left side of Fig. 5.5 illustrates how a classic autoencoder responds when presented with a completely new pattern (active pixels along the secondary diagonal). The autoencoder simultaneously activates Pattern

Fig. 5.2 Left—Activation of the top-left pixel; Right—Output of the trained autoencoder

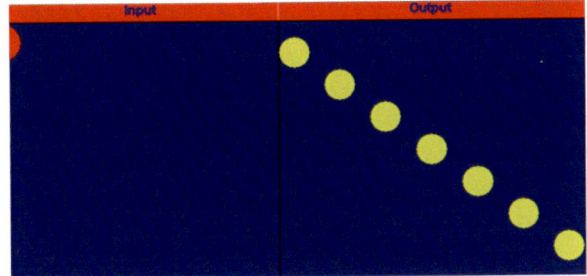

Fig. 5.3 Left—Activation of pixels (2.1) and (3.2); Right—Output of the trained autoencoder

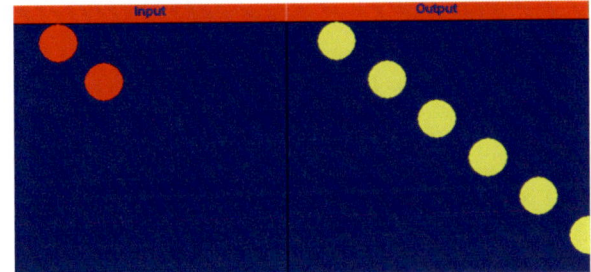

#1 and Pattern #13 from its training dataset as shown in Fig. 5.4c. These are the two patterns it already knows. On the other hand in the right side of Fig. 5.5, the autoencoder responds to a previously unseen pattern by activating training pattern #2, which correctly has the greatest intersection with the new input.

The examples illustrated in Figs. 5.2, 5.3, 5.4 and 5.5 highlight fundamental properties of a trained classic autoencoder:

a. It reconstructs incomplete patterns.
b. When encountering new patterns. it responds with patterns (or combinations thereof) that were already present in its data set according to a logic of vectorial proximity.
c. When responding to new, noisy, or incomplete patterns, it demonstrates that each of its "nodes" disregards the complete structural organization of the dataset on which it was trained; thus. its "knowledge" remains specific and specialized.

We believe that these limitations directly stem from the mathematical techniques employed in training. Both gradient descent and vector quantization are inherently *local techniques*: the former relies on partial derivatives and the latter on cosine angles of similarity between vector pairs (the "winner takes all"). It's akin to the fingers of an imaginary hand initially moving independently and only later (through error backpropagation) learning how their movements synchronized, were in phase, with those of other fingers toward optimizing a goal. Figure 5.6 displays the weight matrices of the trained autoencoder in this toy example (excitations in green. inhibitions in red). It is evident that the matrices lack a global pattern related to the

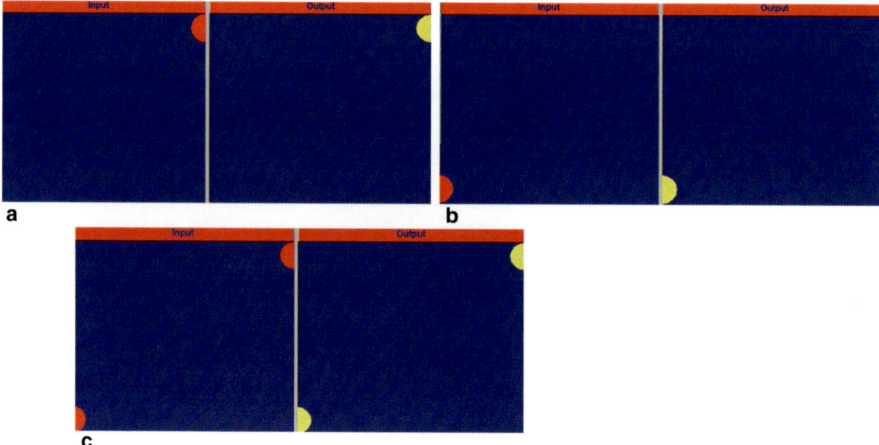

Fig. 5.4 Autoencoder: **a** Activation and output of Pixel (7.1). **b** Activation and output of Pixel (1.7). **c** Activation and output of Pixels (7.1) and (1.7)

Fig. 5.5 Autoencoder: Activation and output when faced with two new input patterns excluded from the training data set

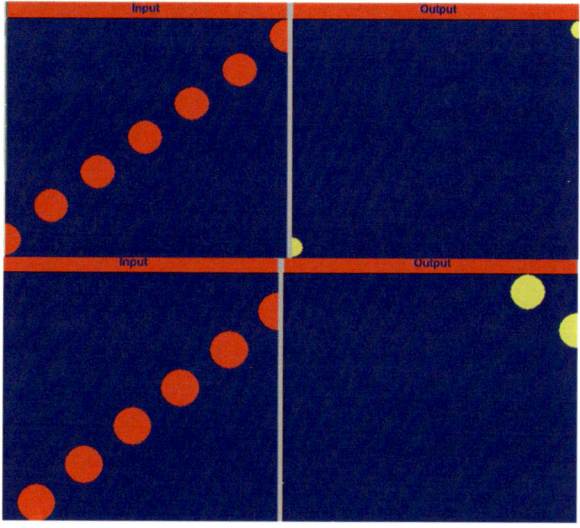

informational structure of the training dataset. Each "weight" resembles an onion encoding different information layers from various input pattern pixels.

However. It is possible to construct autoencoders based on different mathematical principles. For instance. the Auto Contractive Map (Auto CM for short) neural network converges according to the fixed-point theorem and iterative contractions of input patterns though given nonlinearity, it may be a locally convergent point. It captures both specific and global information from the entire training dataset onto its weight matrix (and thus each of its nodes). The architecture of this ANN (see Refs.

Fig. 5.6 Oblique Bars: Weight matrices of the trained autoencoder; (left) Input-Hidden weights, (right) Hidden-Output weights. In red inhibitions; in green excitations between pairs of nodes

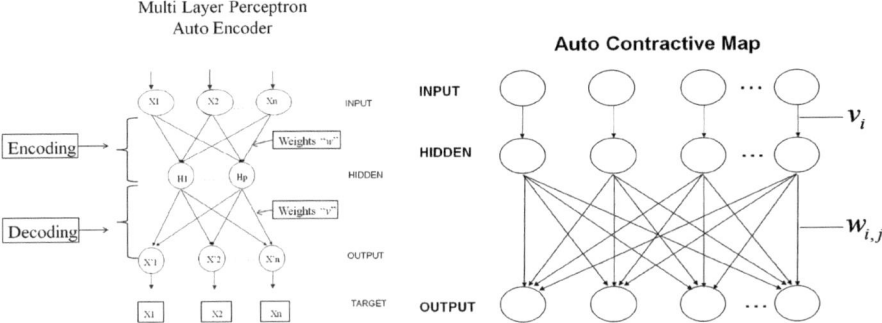

Fig. 5.7 (Left) Topology of a classic autoencoder with a single Hidden layer; (Right) Topology of an Auto Contractive Map (Auto CM)

[1–54]), as shown in Fig. 5.7, isn't vastly different from that of a classic autoencoder, yet its signal propagation and weight correction equations have no relation to gradient descent or error backpropagation concepts.

Significant differences exist between these two types of ANNs in terms of signal flow mathematics, learning equations, and optimizing cost functions. Let's examine the macroscopic distinctions between these two algorithms:

a. Initial weights in the classic autoencoder are random; Auto CM weights are all equal and close to zero (fir example 10^{-6}).
b. The number of input and output nodes in an autoencoder is the same but the number of nodes in intermediate layers (Hidden) is heuristically determined. In contrast. the number of input, hidden, and output nodes in Auto CM is the same as if each variable of an input pattern were written at time #1 (hidden) and time #2 (output).
c. The classic autoencoder corrects its weight matrices through backpropagation (gradient descent and chain rule) of the output error (Output_Vector == Input_Vector) whereas Auto CM autonomously corrects its weight matrices without

Fig. 5.8 Left—Activation
of the top-left pixel;
Right—Output of trained
Auto CM (compare this with
Fig. 5.2)

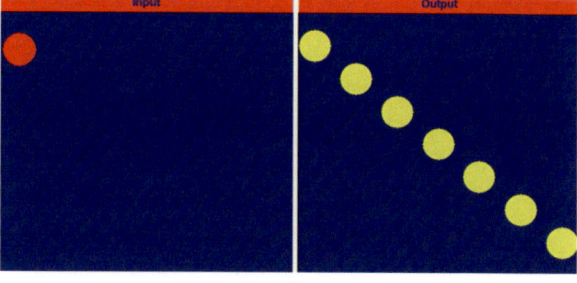

Fig. 5.9 Left—Activation
of pixels (2.1) and (3.2);
Right—Output of trained
Auto CM (compare this with
Fig. 5.3)

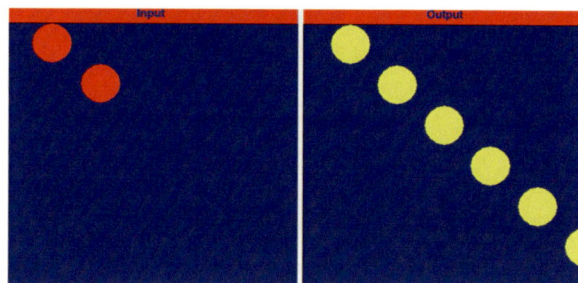

output error backpropagation (Hidden_Vector = = Input_Vector and Output_
Vector = = Hidden_Vector).

d. The cost function characterizing the autoencoder aims to repeat each learning
pattern in the output layer (coding and decoding) while Auto CM optimizes its
cost function when the sum of all its outputs for each input pattern tends towards
zero.

e. At the end of training a classic autoencoder exhibits both excitatory and inhibitory
weights among its nodes whereas Auto CM learns and responds only through
excitatory weights; inhibitory weights do not exist for this ANN.

f. The information in a classic autoencoder after training is distributed across
all trained weight matrices (in the case of a single hidden layer there are two
matrices). In contrast the complete information of a trained Auto CM is only
within the second weight matrix, which fully connects each hidden node to every
output node.

These differences are useful if they lead to significant disparities between the two
types of ANNs. Thus. Let's examine the responses of a trained Auto CM to the same
inputs we provided to a classic autoencoder. Figures 5.8 and 5.9 illustrate that the
two algorithms behave analogously when presented with incomplete patterns.

However, when we feed Auto CM inputs with isolated pixel patterns, the situation
changes drastically (see Fig. 5.10a–c). The activated nodes don't merely display
the most similar patterns on which they were trained (pattern #1 and #7) but they
approximate the global structure of the entire data set.

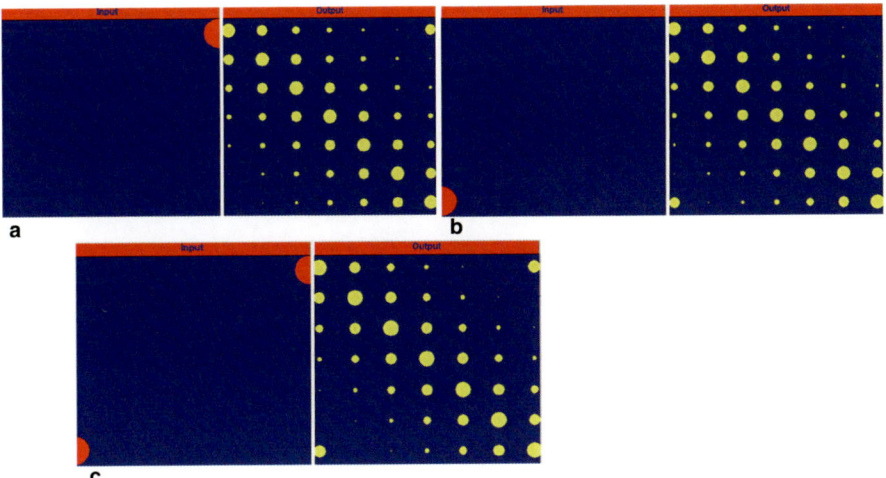

Fig. 5.10 Auto CM: **a** Activation and output of Pixel (7.1). **b** Activation and output of Pixel (1.7). **c** Activation and output of Pixels (7.1) and (1.7) (compare this with Fig. 4.13)

Auto CM's behavior when confronted with completely new patterns is even more surprising and distinct from that exhibited by the classic autoencoder (Fig. 5.11). Auto CM spontaneously convolves the new input patterns presenting the two most similar patterns from its training dataset as output, even if the intersections between the pixels of the two images (input pattern vs. output pattern) are sparse. The explanation for this behavior is rather simple. Auto CM selects output patterns that most closely saturate the input image grid irrespective of the orientation of the utilized "oblique bars".

A more explicit understanding of Auto CM's behavior is gained by observing its weight matrix to which the ANN converges upon the 13 patterns of the toy data set at the end of training (Fig. 5.12). The weight matrix shows a structure that encapsulates the entire training data set through *fractal dynamics*.

Each node of the Auto CM ANN retains both location-specific information and global information about the entire data set. This results is a more robust and efficient behavior when facing new, incomplete, or noisy patterns. Auto CM is an adaptive algorithm operating within the realm of AI whose aim is to uncover hidden information in data. Before delving into a practical example of this on real data, let's provide another toy example showcasing what Auto CM encodes in its weight matrix after training.

Figure 5.13 depicts 9 simple facial expression patterns in an 11 × 11 grid.

Figure 5.14 displays the weight matrices generated by an autoencoder trained on this data set utilizing a single layer of latent units. It is noticeable how these two matrices lack a global pattern approximating the data set's shape. Instead, they are sparse matrices revealing local elementary excitatory and inhibitory features of the 9 learned faces.

Fig. 5.11 Auto CM: Activation and output when faced with two new input patterns excluded from the training dataset

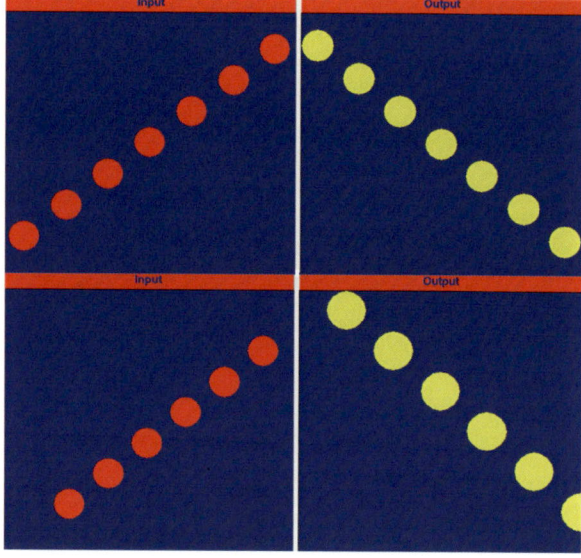

Fig. 5.12 Weight matrix of Auto CM trained on the "oblique bars" data set (green indicates the degree of excitation for each connection)

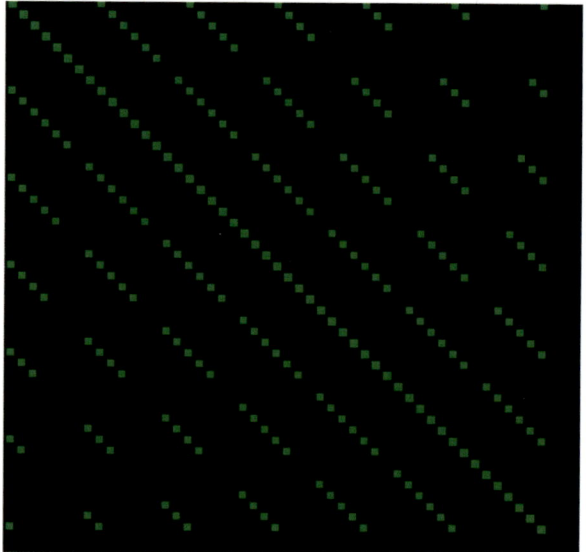

Transforming the same data set using Auto CM generates the weight matrix (121 × 121) shown in Fig. 5.15.

This matrix (Fig. 5.15) warrants a few observations:

a. The entire informational structure of the data set is projected onto Auto CM's weight matrix.

Fig. 5.13 9 schematic facial
expression patterns in an 11
× 11 square

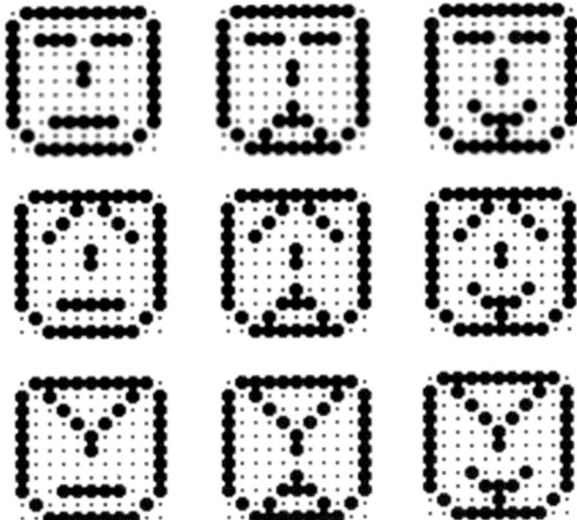

b.   The elements comprising the global face structure are in themselves facets (fractal structure).

c.   Each elementary facet, composed of 11 × 11 pixels, is characterized by its position in the global face. It's as if each facet adapted and transformed itself based on the position it occupies within the larger face of which it's a component.

5.1   Application on Real Data: The Gang and the Two Clans: The Data and the Problem

Police investigation data is generally incomplete and often contradictory. Learning from a large amount of data from different investigations of the same type of crime does not solve the problem. Each investigation has its own characteristics, attributes, and events which are often few and scattered. AI useful for law enforcement must provide precise analysis for each specific investigation not just a general statistical prediction of crimes of a certain type in certain places. The AI presented below is tasked with showing this type of precision analysis: a specific investigation, with specific characters characterized by attributes collected often randomly and unsystematically ("XX cousin of YY", "ZZ son of QQ", "KK has a car with license plate CC", "BB appears in the same document with NN and WW", and so on).

Fig. 5.14 Faces: Weight matrices of the trained autoencoder; (left) input-hidden weights. (right) hidden-output weights. In red, inhibitions; in green, excitations between pairs of nodes

The investigation we present[1] took place between 2008 and 2009. and it consists of 62 characters (whose names have been replaced by numbers for confidentiality) monitored from the beginning of the investigations by law enforcement agencies. The attributes characterizing each of these characters can be grouped into 12 macro variables (see Fig. 5.16).

---

[1] On July 9, 2008, the Commander of the ROS approved the research project managed by Semeion and requested the deployment of the Auto CM system, authorized by the III Department—CGA; On July 14, 2008, the operational phase was initiated; On October 28, 2008, a training and update session for the personnel of the ROS—Technical Investigations and Analysis Units took place at the SEMEION Research Center in Rome.

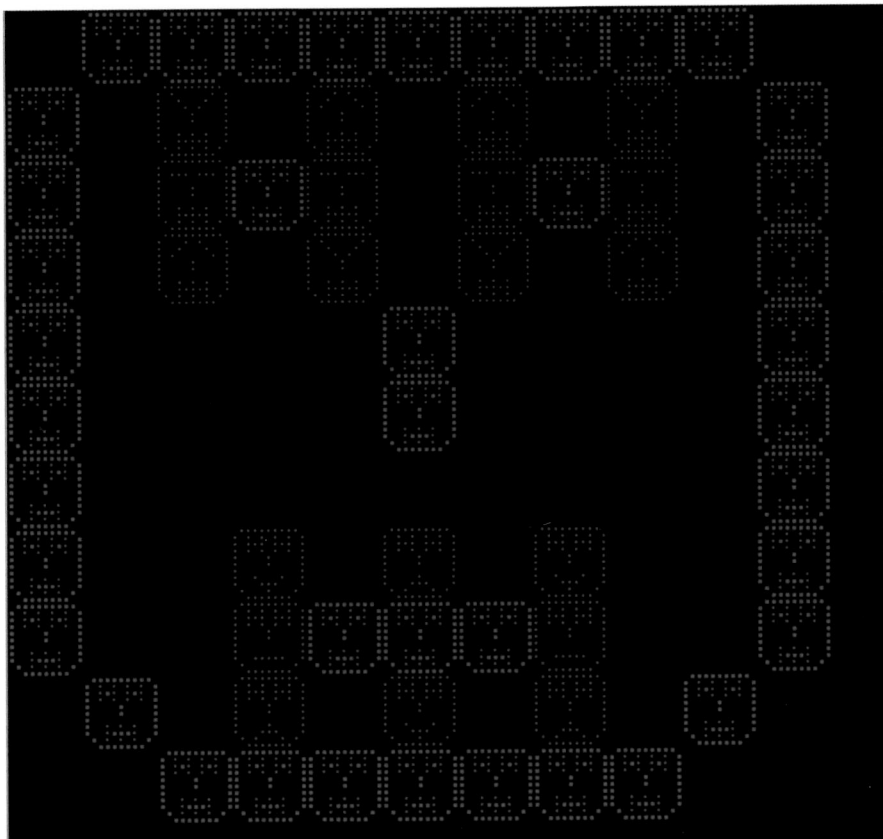

Fig. 5.15 Auto CM: Weight matrix (121 × 121) after training on the "Faces" dataset

Num	Macro Variable	Meaning
1	PERSON_RELATIONSHIP	With which individuals does one have familial ties
2	CLOSE_ACQUAINTANCE	With which individuals does one have frequent interactions
3	FRIENDSHIP_INTERACTION	With which individuals does one have friendship and social interactions
4	WORK_BUSINESS_ASSOCIATES	With which individuals does one have work-related affairs
5	LEGAL_CONNECTION	With which individuals has one been accused and/or implicated
6	CRIMINAL_COHORTS	With which individuals is one an accomplice in criminal actions
7	VICTIMS	Which individuals are known to be victims
8	PROPERTY_REAL_ESTATE	What properties does one own partially or wholly
9	CORPORATIONS	In which companies is one involved in any capacity
10	ADMINISTRATIVE_DOCUMENTS	In which administrative documents is one mentioned
11	TELEPHONE_USAGE	Which telephone lines are registered to one
12	VEHICLES	Which vehicles does one own

Fig. 5.16 Macro variables available to characterize each of the 62 subjects

These macro variables have been expanded so that each macro variable was rewritten into a number of micro variables equal to the number of its options. Each micro variable in this way could take a Boolean value (True = 1 vs. False = 0). Such pre-processing led to a significant increase in the variables of the data set. which became 778 columns (micro variables) and 62 rows (subjects under attention).

However, even with this rewriting, the data set was shown to be very poor. Traditional statistical analyses had proved futile in revealing information that investigators considered new or relevant and the investigations were on the verge of being closed after a series of halts based on evidence and clues that emerged during the investigations themselves. It should be added that during the investigations, the investigators had discovered that the 62 subjects under attention were distributed in two clans. This information was not present in the original data set (Fig. 5.16) and therefore it could be used as a "gold standard" to evaluate the different algorithms to be applied to the base data set, which had so far been of little use. If at least one of the adaptive algorithms applied to the data set had spontaneously separated the subjects of one clan from those of the other with a certain accuracy, then that algorithm could be used to provide new insights to the investigations. Figure 5.17 shows how the 62 subjects were divided into the two clans which we will call "red clan" and "blue clan" for simplicity.

5.2   The Gang and the Two Clans: The Method

A data set consisting of 62 observations (subjects under attention) and 778 Boolean variables, {1.0}, is difficult to analyze with probabilistic tools since the number of "zeros" is very high (see Fig. 5.18). For this reason, three types of algorithms have been chosen to conduct this analysis and listed below.

a. **The Manifest Physical Network**: This is a very simple algorithm that reconstructs the "physical network" of connections between each subject under attention and every other. This will result in a square and symmetrical matrix of 62 columns by 62 rows in which for each subject, the number of variables shared with every other will be counted (for example, vehicle and phone sharing, family ties, participation in the same societies, property co-ownership, and so on). This explains information that is already explicitly contained in the data set. Therefore, no AI technique is implied in this approach. We define the matrix produced by this algorithm as the "Evident Matrix".

b. **The Second-Order Hidden Network**: Various types of autoencoders have been used for this analysis, some shallow (a classic MLP with only one layer of latent units), others deep such as variational autoencoders (VAEs:—see Refs. [54, 55]) and new recirculation ANNs (NRC:—see Ref. [56]) with up to 6 intermediate layers between input and output. Each of these ANNs generated multiple matrices of tensors associating each subject to every other based on the similarity of the variables characterizing each one. The various ANNs used were measured based

Fig. 5.17 The 62 subjects under attention divided into two clans (red and blue), information discovered during the investigations and not present in the analysis of the data set (the light blue color is for a subject connected to both clans)

Subjects				
ID	**Name**		**ID**	**Name**
1	80		32	3150
2	1389		33	3163
3	1391		34	3447
4	1662		35	4861
5	1827		36	4868
6	1925		37	4869
7	1927		38	4870
8	1928		39	4871
9	1929		40	4872
10	1931		41	4879
11	1932		42	4888
12	1933		43	5803
13	1973		44	6226
14	2360		45	7706
15	2513		46	22435
16	2515		47	34513
17	2516		48	34514
18	2517		49	35238
19	2526		50	36261
20	2532		51	39265
21	2534		52	39266
22	2536		53	39267
23	2538		54	39270
24	2543		55	39271
25	2553		56	39272
26	2555		57	39273
27	2934		58	39274
28	2949		59	39275
29	2958		60	39276
30	3082		61	39278
31	3085		62	39279

on their ability to blindly distinguish the membership of each subject to its clan (Red Clan vs. Blue Clan) and the best ANN was selected. For this case the final output of each ANN is also a 62×62 matrix where the similarity between each subject and every other is measured. We refer to these as second-order ANNs because the similarity between each subject and every other is measured considering all the variables of each one simultaneously.

c. **The Third-Order Hidden Network:** The algorithm used for this third method was Auto Contractive Map (Auto CM: see Ref. [36])), a particular ANN already presented. Auto CM is capable of analyzing the relationships between subjects and variables up to the 3rd order and generates a single matrix of tensors in the form of a 62×62 matrix.

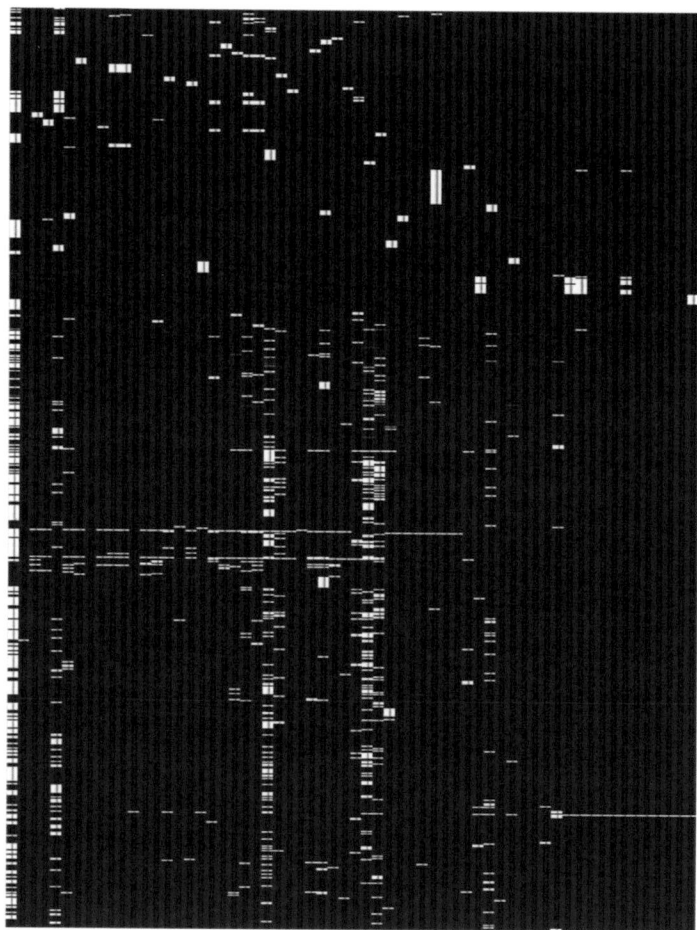

Fig. 5.18 The data set of 778 rows (variables) and 62 columns (subjects under attention): in black the variables with value "0", in white the variables with value "1"

All the methods used provide an output in the form of a 62 × 62 matrix where the connection and/or similarity between all the subjects under attention are measured. To make the many meanings of these matrices comprehensible and also visual, each of them has been filtered through a particular graph called the Minimum Spanning Tree (MST). This graph has special properties both in mathematics and physics (see Refs. [58, 59]). The MST shows a system in its conditions of minimum potential, where the system tends to over time if the conditions at its boundaries remain more or less stable. The MST, the relationship of connection and/or similarity between the 62 subjects of the investigation according to the 778 attributes with which each of them was characterized.

A limitation of the MST is that it's a tree graph, that is, without the possibility of creating circuits between its nodes (subjects under attention) and always having to find a connection between all the subjects even if it's very weak. To overcome the first of these inconveniences, which often eliminates important connections between subjects just because they create cliques, a new type of graph called the *Maximally Regular Graph* (MRG: see Ref. [36])) is introduced. The MRG adds the most important circuits that emerge from the various nodes to the MST. Once the various MSTs were extracted from the matrices of the three methods, the connections between subjects of the same clan and subjects of the opposite clan predicted by each MST were analyzed. This was done to verify how well the MSTs of different ANNs, and algorithms spontaneously distinguished the two clans whose existence had only been discovered during the investigations. The criterion by which the accuracy of each MST is calculated is simple. If P is the Number of Classes-1, N is the Number of Links in the MST, and E(i. j) is the error of connection between two nodes belonging to different classes, then the accuracy of each MST will be given by the following equation:

$$MST\_Accuracy = 1.0 - \frac{\sum_{i}^{N-1} \sum_{j=i+1}^{N} E_{i,j} - P}{N \cdot (N-1)}.$$

This equation takes into account various factors:

a. The MST always connects all the nodes of the graph in a tree-like path so if the classes to be distinguished are 2, then 1 connection will necessarily be wrong. If the classes were 3, the necessarily wrong connections would be 2, and so on, In fact, P-NumClasses-1 will be wrong.
b. The MST always predicts a number of links equal to the number of its nodes-1. Thus L is the number of NODES-1.
c. The MST predicts symmetric connections and does not predict self-connection so that the number of links to consider will be equal to N(N-1).

The measure we have defined, *MST_Accuracy*, shows how much the MST graph generated from the similarity matrix of a specific algorithm correctly groups/clusters the classes. That is, MST_Accuracy measures how well the examined algorithm is able to infer, solely from the information contained in the input patterns (the 778 variables of the 62 subjects under attention in the data set under examination), to which class (clan) each subject belongs.

Figure 5.19a contains the results of the algorithms we have compared. while in Fig. 5.19b, the same information appears in the form of a graph.

We illustrate the MST of the top 3 algorithms In the following figures, Figs. 5.20, 5.21 and 5.22. Even visually. the scores assigned in Fig. 5.19a are evident.

The MRG of the Auto CM ANN was generated to explicitly reveal any implicitly present cliques in the graph that were not detectable by the simple MST based on these results (see Fig. 5.23).

It is clear from this graph, that the Red Clan and the Blue Clan are organized differently. If it is true that the morphology of a structure of the graph also informs about its possible functioning, the Blue Clan has its parliament consisting of a diamond where all members interact. The Red Clan, on the other hand, seems more scattered and

Algorithm	Graph Accuracy	Graph Errors	Explicit Algorithm Name
(Auto CM)	0.967213	2	Auto Contractive Map
(L3)(NRC)	0.868852	8	New Recirculation ANN with 6 Hidden Layers
Psysical Networks	0.819672	11	Blatant Physical Networks
(L5)(NRC)	0.819672	11	New Recirculation ANN with 10 Hidden Layers
(L1)(VAE)	0.770492	14	Variational Auto encoder with 1 Latent Layer
(L3)(VAE)	0.770492	14	Variational Auto encoder with 1 Latent Layer and two Hidden Layer
(L5)(VAE)	0.770492	14	Variational Auto encoder with 1 Latent Layer and three Hidden Layers
(L1)(C-BP)	0.754098	15	Classic Back Propgation Autoencoder with 1 Hidden Layer
(L3)(C-BP)	0.688525	19	Classic Back Propgation Autoencoder with 3 Hidden Layers

a

Graph Accuracy

[bar chart with y-axis values 0.5, 0.55, 0.6, 0.65, 0.7, 0.75, 0.8, 0.85, 0.9, 0.95, 1 and x-axis categories (Auto CM), (L3)(NRC), Psysical Networks, (L5)(NRC), (L1)(VAE), (L3)(VAE), (L5)(VAE), (L1)(C-BP), (L3)(C-BP)]

b

Fig. 5.19 **a** Comparison of MST Accuracy in the tested algorithms on the Gang data set (62 × 778). **b** Histogram of MST Accuracy in the tested algorithms on the Gang data set (62 × 778). In red the best algorithm, in green the classical one that displays the relationships already present in the database

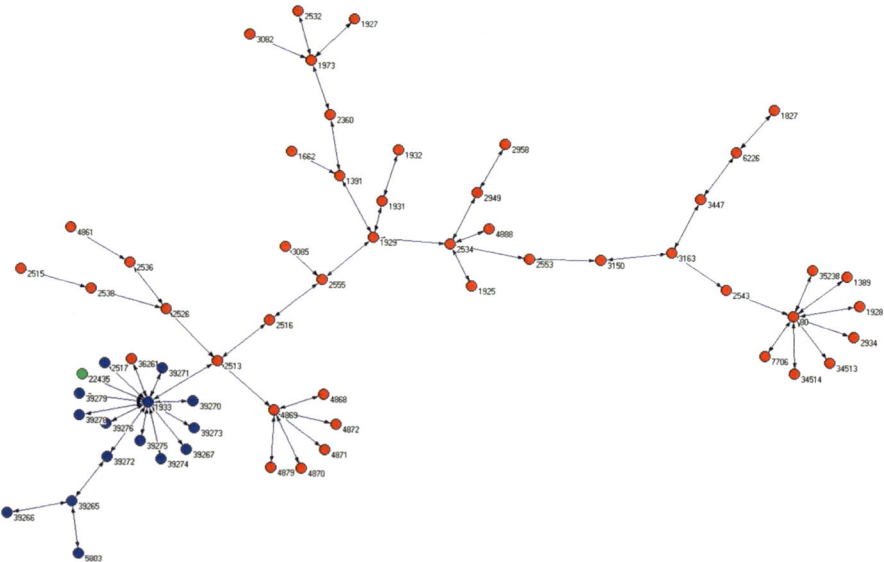

Fig. 5.20 MST Graph of the Auto Contractive Map (Auto CM) ANNs. The colors of the two clans were added after the results of the processing. The algorithm executed its processing without this information. There are only 2 errors. These 2 could be 1 considering that Subject 22,435 in green belongs to both clans

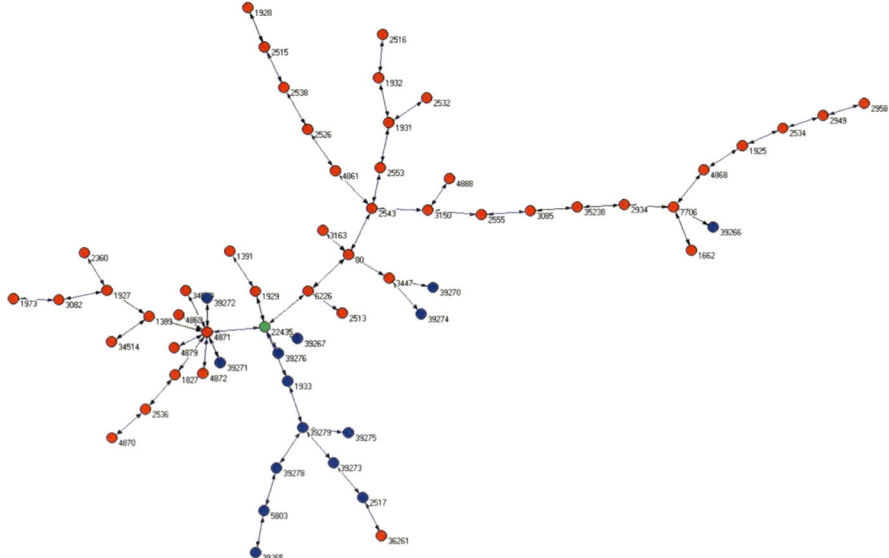

Fig. 5.21 MST Graph of the New Recirculation (L3 NRC) ANNs. The colors of the two clans were added after the results of the processing. The algorithm executed its processing without this information. There are 6 errors that could be 5 considering that Subject 22,435 in green belongs to both clans

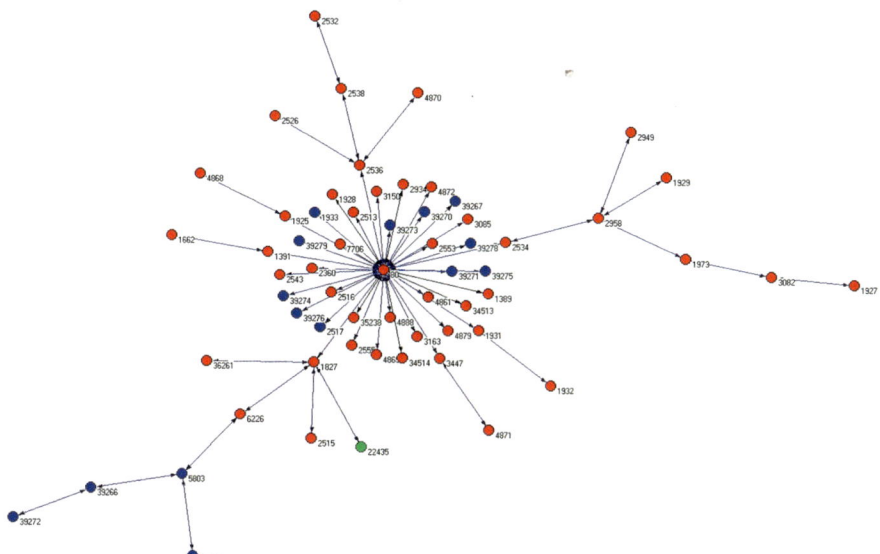

Fig. 5.22 MST Graph according to the Physical Networks. The colors of the two clans were added after the results of the processing. The algorithm executed its processing without this information. There are only 11 errors that could be 10 considering that Subject 22,435 in green belongs to both clans

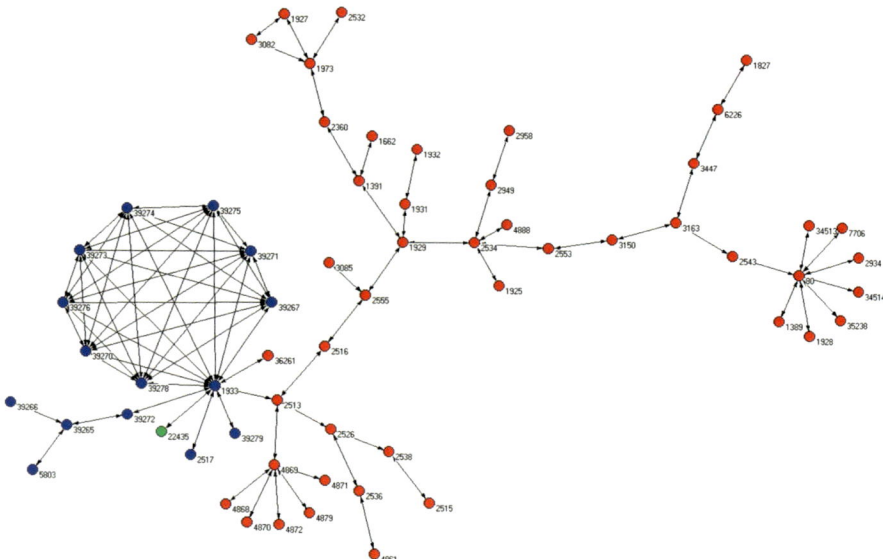

Fig. 5.23 MRG [35, 57] generated by Auto CM: two cliques become visible in the graph, one mostly involving the Blue Clan and the other a small relational triangle within the Red Clan

less robust. These inferences would have to be confirmed and/or refuted by further investigations carried out by the investigators. However, the most important information to extract from this graph concerns the importance that each member of each of the two Clans has in holding the entire gang together.

The traditional indices used in network analysis (betweenness, degree, strength, and so on) do not help in this analysis. The leaves of the graph, far from representing the less important members of the gang could simply be generated by the type of variables considered. For example, leaves are infrequent but perhaps as important as outliers are in traditional statistical analyses. This being the case, a new type of analysis was carried out consisting of two steps:

a. Remove a member of the data set from the tensor matrix and generate a graph without that member.
b. Compare the complexity of the original graph with the complexity of the one obtained without a gang member.

This comparison can show how indifferent the removed member is to the gang's structure. Is the original complexity similar to the one without that member? How much does the individual gang member's presence increases the structural complexity of the gang (the original complexity greater than the one)? Or how much does their presence inhibit the growth of the gang (original complexity smaller than the one without that member)? The measure used for analyzing the complexity of the graph is called the H0 Index (see Ref. [59]). Figure 5.24 shows the same graph as Fig. 5.23

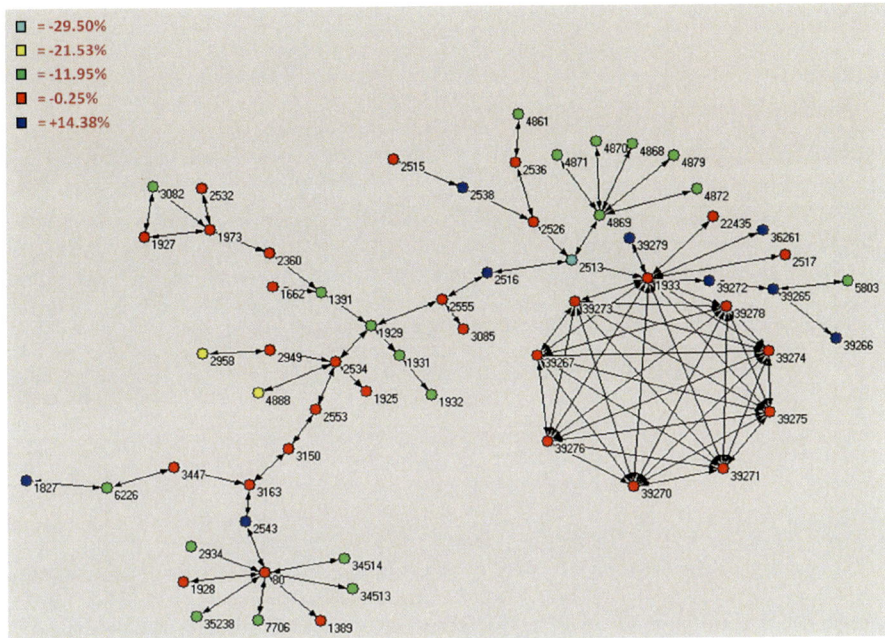

Fig. 5.24 Contribution of each gang member to the robustness of the graph. Subjects in light blue and yellow are those whose absence causes the greatest weakening of the graph (see the legend)

but this time the colors of each node (subject) are connected to the positive or negative neuro-contribution that node offers to the graph's complexity.

It is clear, from Fig. 5.24, that according to Auto CM, Subject 2513 (celeste color) is the most important for maintaining the complexity of the gang. The absence of Subject 2513 (arrest) could lead to a simplification of the network of the entire Gang by almost 30%. Other important subjects for maintaining the gang's complexity are Subjects 4888 and 2958 (in yellow) whose absence would cause a collapse of 21% in the complexity of the entire structure. On the other hand, Subject 80 (in red), defined by the investigators as the "leader" of the gang, is considered by Auto CM as a member whose absence would not modify the gang's structure.

5.3 Results from Further Investigations

We now present some official observations that the investigators wrote after becoming aware of the hypotheses put forward by the Auto CM ANN and the MRG graph and after conducting further investigations[2] (extracts from the 'Project Magellano:

---

[2] All the citations from the carabinieri official report are in *italic*.

Taurania Investigation' by the Technical Investigation Unit of ROS, data extracted from the SYNAPSIS system of ROS, December 12, 2008).

> The info-investigative data of the investigation had already been entered into the Synapsis system as 'entities'. The data pertains to 1112 individuals; 2 organizations; 118 companies; 68 phone lines; 258 vehicles; 475 addresses; 353 documents; 9583 relationships among the entities. (Taurania Investigation slide #11).
>
> From an initial analysis of the variables (attributes or characteristics), it was possible to outline a graph that relates individuals based on their affinities, defining a possible hierarchy in the organizational structure of the two clans. (Taurania Investigation slide #18).
>
> The analysis, carried out with a proprietary algorithm SEMEION (MRG-AutoCM), allowed us to identify individuals who, within the two different organizations, are:
>
> a. *more protected within their own organizational structure (mafia-type associations or drug trafficking-related);*
> b. *more influential within the graph (which encompasses the two organizations).*
>
> Hypothetically, the 'damage' that would be caused to the structure by subtracting the functional contribution of each individual was thus calculated. i.e.. the energy that—in that case—the structure itself requires to reorganize. (Taurania Investigation slide #19).
>
> The most protected individual is found to be id 2513 (Name redacted, see Fig. 5.24, editor's note), an individual considered to be very close to clan leader XXX.XXX (Red Clan, editor's note) and connected to other subjects of investigative interest. The analysis highlighted an important position whose significance had not emerged during the investigation. (Taurania Investigation slide #20).
>
> An individual of high info-investigative interest is represented by id 4888 (Name redacted, see Fig. 5.24), leader of the Name Redacted clan allied with the XXX.XXX clan (Red Clan, editor's note), but remaining on the sidelines of the investigative activity due to his 'transversal' position. Id 4888 manages relationships with other Camorra organizations. (Taurania Investigation slide #21).
>
> A second individual of info-investigative interest is represented by id 2958 (Name redacted, see Fig. 5.24). a drug trafficker, supplier to some clans of the Camorra Vesuviana and the city of Naples, with connections to Colombian suppliers. He is not under investigation due to investigative coordination needs. (Taurania Investigation slide #21).
>
> The individuals marked with the blue sign are the so-called 'conservatives', who keep the structure closed. Their absence could even facilitate opening and connections to the outside. The four individuals hit by O.C.C. (Preventive Custody Order) are indicated with the blue circle (Fig. 5.25). (Taurania Investigation slide #22).

This analysis with Auto CM was subsequently repeated considering not only the initial 62 subjects but all 1112 subjects that appeared in the investigation. The variables and attributes characterizing each of the new subjects had obviously increased to become 8586. Figure 5.26 shows the graph of the 1112 subjects (including the previous 62). This further analysis aimed to understand which subjects beyond the initial 62, played an important role in the organization of the gang.

Here are some findings from the investigators:

> Id 22,550, Name Redacted (whose father had been the victim of an attack), arrested with O.C.C., was associated with the Red Clan. but in the SYNAPSIS information system of the ROS, he had not been cataloged among the 62 associates. **Auto CM positioned him correctly in the Red Clan** (Indagine Turania slide #38).

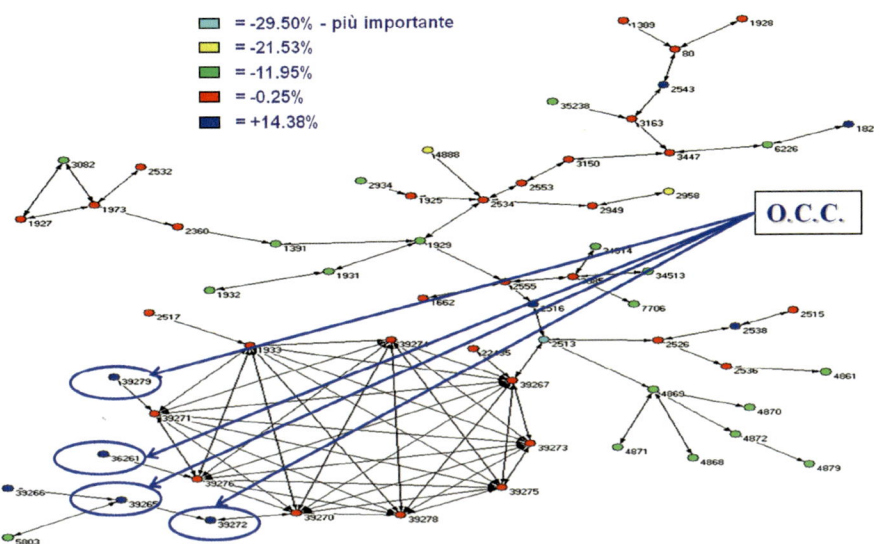

Fig. 5.25 Four of the individuals who, according to Auto CM, kept the gang closed, affected by the Preventive Custody Order

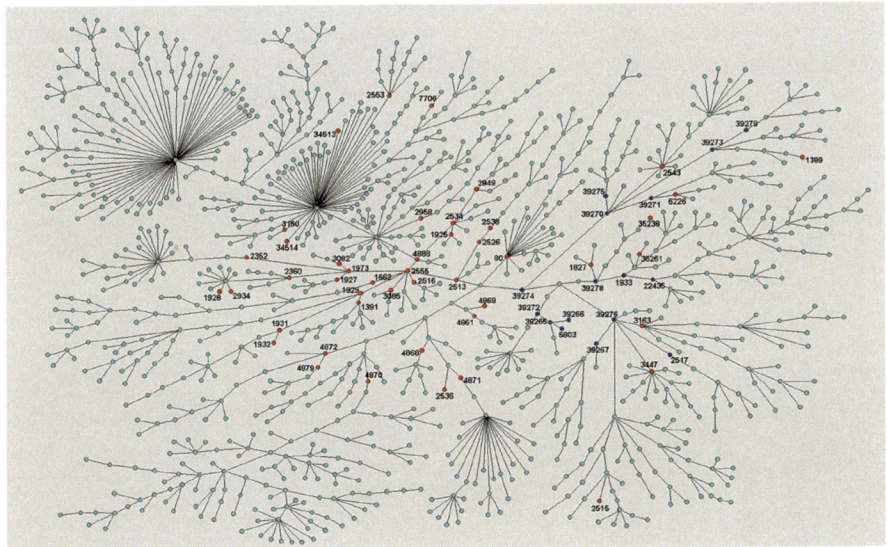

Fig. 5.26 MST of the 1112 subjects after the analysis with Auto CM. Marked in red and blue are the 62 previously analyzed

Id 6120, Name Redacted, arrested with O.C.C., previously not included in either of the two clans, **is placed by Auto CM in the Red Clan**. (Indagine Turania slide #38).

Id 4873, Name Redacted, is in the center of the graph and is therefore—for the analysis—the most protected person. **He was hit by O.C.C., in which he is indicated as "constantly available to the clan leader for the illicit purposes of the organization"** (Indagine Turania slide #33).

Id 6189, Name Redacted, who is found to be most similar to clan leader id 80 according to Auto CM, **is the main facilitator** in whose residence at the end of September 2008 fugitive id 80 was arrested (Indagine Taurania slide #34).

5.4   Conclusions

a. *The analysis is validated by the investigative outcomes;*
b. *The tool has proven to be of high utility in tactical analysis to guide investigative choices;*
c. *The system can also be used in operational analysis on criminal contexts of interest." (Indagine Turania slide #35).*

The application of the Auto CM ANN to a real and solved case with the help of this technology not only demonstrates how an ANN whose nodes consciously and cooperatively learn both a specific part of the input data and the entire structure of the dataset on which they are trained is more effective than an ANN where each node remains unaware of what the other nodes of the same ANN are learning. It also shows that only ANNs whose nodes cooperate holistically in learning a problem (data set) are capable of revealing non-trivial hidden information within that data structure. Therefore, what has been presented aligns precisely with the direction of an AI whose approach is predominantly holistic and we have defined as a investigative approach.

References

1. Buscema M (2007) A novel adapting mapping method for emergent properties discovery in data bases: experience in medical field. In: 2007 IEEE international conference on systems, man and cybernetics (SMC 2007). Montreal, Canada, 7–10 Ottobre 2007
2. Buscema M., Grossi E (2008) The semantic connectivity map: an adapting self-organizing knowledge discovery method in data bases. Experience in gastro-oesophageal reflux disease. Int J Data Min Bioinf 2(4)
3. Buscema M, Grossi E, Snowdon D, Antuono P (2008) Auto-contractive maps: an artificial adaptive system for data mining. An Appl Alzheimer Dis Curr Alzheimer Res 5:481–498
4. Buscema M (ed) (2007) Squashing theory and contractive map network, semeion technical paper #32, Rome
5. Buscema M, Helgason C, Grossi E (2008) Auto contractive maps, H function and maximally regular graph: theory and applications. In: Special session on "artificial adaptive systems in medicine: applications in the real world, NAFIPS 2008 (IEEE), New York, 19–22 May 2008
6. Licastro F, Porcellini E, Chiappelli M, Forti P, Buscema M et al (2010) Multivariable network associated with cognitive decline and dementia. Int Neurobiol Aging 1(2):257–269

7. Buscema M, Grossi E (eds) (2009) Artificial adaptive systems in medicine, Bentham e-books, pp 25–47

8. Buscema M, Sacco PL (2010) Auto-contractive Maps, the H Function, and the maximally regular graph (MRG): a new methodology for data mining. In: Capecchi V et al (eds) Applications of mathematics in models, artificial neural networks and arts, Chapter 11. Springer Science+Business Media B.V. https://doi.org/10.1007/978-90-481-8581-8_11

9. Grossi E, Blessi GT, Sacco PL, Buscema M (2011) The interaction between culture, health and psychological well-being: data mining from the Italian culture and well-being project. J Happiness Stud

10. Licastro F, Porcellini E, Forti P, Buscema M, Carbone I, Ravaglia G, Grossi E (2010) Multi factorial interactions in the pathogenesis pathway of Alzheimer's disease: a new risk charts for prevention of dementia. Immunity Ageing 7(Suppl 1):S4

11. Buscema M, Newman F, Grossi E, Tastle W (2010) Application of adaptive systems methodology to radiotherapy. In: NAFIPS 2010, 12-14 July, Toronto, Canada

12. Eller-Vainicher C, Zhukouskaya VV, Tolkachev YV, Koritko SS, Cairoli E, Grossi E, Beck-Peccoz P, Chiodini I, Shepelkevich AP (2011) Low bone mineral density and its predictors in Type 1 diabetic patients evaluated by the classic statistics and artificial neural network analysis. In: Diabetes care, pp 1–6

13. Gomiero T, Croce L, Grossi E, De Vreese L, Buscema M, Mantesso U, De Bastiani E (2011) A short version of SIS (support intensity scale): the utility of the application of artificial adaptive systems. US-China Educ Rev A 2:196–207

14. Buscema M, Penco S, Grossi E (2012) A novel mathematical approach to define the genes/SNPs conferring risk or protection in sporadic amyotrophic lateral sclerosis based on auto contractive map neural networks and graph theory. Neurol Res Int 2012, Article ID 478560, 13 pp. https://doi.org/10.1155/2012/478560

15. Grossi E, Compare A, Buscema M (2012) The concept of individual semantic maps in clinical psychology: a feasibility study on a new paradigm. Q & Q Int J Methodol. ISSN 0033-5177, Qual Quant. https://doi.org/10.1007/s11135-012-9746-8

16. Coppedè F, Grossi E, Buscema M, Migliore L (2013) Application of artificial neural networks to investigate one-carbon metabolism in alzheimer's disease and healthy matched individuals. PLOS ONE 8(8):e74012, 1–11. www.plosone.org

17. , Street ME, Buscema M, Smerieri A, Montanini L, Grossi E (2013) Artificial neural networks, and evolutionary algorithms as a systems biology approach to a data-base on fetal growth restriction. Prog Biophys Mol Biol 1–6

18. Compare A, Grossi E, Buscema M, Zarbo C, Mao X, Faletra F, Pasotti E, Moccetti T, Mommersteeg PMC, Auricchio A (2013) Combining personality traits with traditional risk factors for coronary stenosis: an artificial neural networks solution in patients with computed tomography detected coronary artery disease. Cardiovascul Psychiat Neurol, Article ID 814967, 9 pp. https://doi.org/10.1155/2013/814967

19. Buscema M, Consonni V, Ballabio D, Mauri A, Massini G, Breda M, Todeschini R (2014) K-CM: a new artificial neural network. Application to supervised pattern recognition. Chemometr Intell Lab Syst 138(2014):110–119

20. Buscema M, Massini G, Maurelli G (2014) Artificial neural networks: an overview and their use in the analysis of the AMPHORA-3 dataset. Substance Use Misuse, Early Online:1–14

21. Gironi M, Borgiani B, Farina E, Mariani E, Cursano C, Alberoni M, Nemni R, Comi G, Buscema M, Furlan R (2015) And Enzo Grossi, a global immune deficit in alzheimer's disease and mild cognitive impairment disclosed by a novel data mining process. J Alzheimer's Disease 43:1199–1213

22. Drenos F, Grossi E, Buscema M, Humphries SE (2015) Networks in coronary heart disease genetics as a step towards systems epidemiology. PLoS ONE 10(5):e0125876. https://doi.org/10.1371/journal.pone.0125876

23. Coppedè F, Grossi E, Lopomo A, Spisni R, Buscema M, Migliore L (2015) Application of artificial neural networks to link genetic and environmental factors to DNA methylation in colorectal cancer. Epigenomics 7(2):175–186

24. Narzisi A, Muratori F, Buscema M, Calderoni S, Grossi E (2015) Outcome predictors in autism spectrum disorders preschoolers undergoing treatment as usual: insights from an observational study using artificial neural networks. Neuropsychiatric Disease Treat 11:1587–1599
25. Buscema M, Grossi E, Montanini L, Street ME (2015) Data mining of determinants of intrauterine growth retardation revisited using novel algorithms generating semantic maps and prototypical discriminating variable profiles. PLoS ONE 10(7):e0126020. https://doi.org/10.1371/journal
26. Buscema PM, Gitto L, Russo S, Marcellusi A, Fiori F, Maurelli G, Massini G, Mennini FS (2016) The perception of corruption in health: AutoCM methods for an international comparison. Qual Quant. https://doi.org/10.1007/s11135-016-0315-4
27. Ferilli G, Sacco PL, Teti E, Buscema M (2016) Top corporate brands and the global structure of country brand positioning: an AutoCM ANN approach. Exp Syst Appl 66:62–75.
28. Buscema M, Sacco PL (2016) MST fitness index and implicit data narratives: a comparative test on alternative unsupervised algorithms. Physica A 461:726–746
29. Buscema M, Ferilli G, Sacco PL (2017) What kind of 'world order'? An artificial neural networks approach to intensive data mining. Technol Forecast Soc Change 117:46–56
30. Mennini FS, Gitto L, Russo S, Americo, Cicchetti, Ruggeri M, Coretti S, Maurelli G, Buscema PM (2017) Does regional belonging explain the similarities in the expenditure determinants of Italian healthcare deliveries? An approach based on artificial neural networks. Econ Anal Policy 55:47–56
31. Buscema M, Ferilli G, Sacco PL (2017) What kind of 'world order'? An artificial neural networks approach to intensive data mining. Technol Forecast Soc Change 117:46–56
32. Friedel MJ, Buscema M, Vicente LE, Iwashita F, Koga-Vicente A (2017) Mapping fractional landscape soils and vegetation components from Hyperion satellite imagery using an unsupervised machine learning workflow. Int J Dig Earth. https://doi.org/10.1080/17538947.2017.1349841
33. Buscema M, Ferilli G, Sacco PL (2018) The meta-geography of the open society: an Auto-CM ANN approach. Exp Syst Appl 99:12–24
34. Vigna L, Brunani A, Brugnera A, Grossi E, Compare A, Tirelli AS, Conti DM, Agnelli GM, Andersen LL, Buscema M, Riboldi L (2018) Determinants of metabolic syndrome in obese workers: gender differences in perceived job-related stress and in psychological characteristics identified using artificial neural networks. Eat Weight Disorders Stud Anorexia Bulimia Obesity. https://doi.org/10.1007/s40519-018-0536-8
35. Buscema M, Massini G, Breda M, Lodwick W A, Newman F, Asadi-Zeydabadi M (2018) Artificial adaptive systems using auto contractive maps—theory, applications and extensions. Stud Syst Decis Control 131. https://doi.org/10.1007/978-3-319-75049-1. ISBN 978-3-319-75048-4
36. Grossi E, Podda GM, Pugliano M, Gabba S, Verri A, Carpani G, Buscema M, Casazza G, Cattaneo M (2014) Prediction of optimal warfarin maintenance dose using advanced artificial neural networks. Pharmacogenomics 15(1):29–37
37. Buscema M, Consonni V, Ballabio D, Mauri A, Massini G, Breda M, Todeschini R (2014) K-CM: a new artificial neural network, application to supervised pattern recognition. Chemometr Intell Lab Syst 138:110–119
38. Buscema PM, Massini G, Maurelli G (2014) Artificial neural networks an overview and their use in the analysis of the AMPHORA-3 dataset. Substance Use Misuse Early Online:1–14
39. Gironi M, Borgiani B, Farina E, Mariani E, Cursano C, Alberoni M, Nemni R, Comi G, Buscema M, Furlan R, Grossi E (2015) Alzheimer's disease and mild cognitive impairment disclosed by a novel data mining process. J Alzheimer's Disease 43:1199–1213. https://doi.org/10.3233/JAD-141116
40. Vigna L, Tirelli AS, Grossi E, Turolo S, Tomaino L, Napolitano F, Buscema M, Riboldi L (2019) Directional relationship between vitamin D status and prediabetes: a new approach from artificial neural network in a cohort of workers with overweight-obesity. J Am Coll Nutr. https://doi.org/10.1080/07315724.2019.1590249

41. Buscema M, Ferilli G, Gustafsson C, Sacco PL (2019) The complex dynamic evolution of cultural vibrancy in the Region of Halland, Sweden. Int Reg Sci Rev. https://doi.org/10.1177/0160017619849633

42. Brancato A, Buscema PM, Massini G, Gresta S (2016) Pattern recognition for flank eruption forecasting: an application at Mount Etna Volcano (Sicily, Italy). Open J Geol 6:583–597

43. Brancato A, Buscema PM, Massini G, Gresta S, Salerno G, Torre FD (2019) K-CM application for supervised pattern recognition at Mt. Etna: an innovative tool to forecast flank eruptive activity. Bull Volcanol 81(40). https://doi.org/10.1007/s00445-019-1299-4

44. Bronzi B, Brilli C, Beone GM, Fontanella MC, Ballabio D, Todeschini R, Consonni V, Grisoni F, Parri F, Buscema M (2020) Geographical identification of Chianti red wine based on ICP-MS element composition. Food Chemi. https://doi.org/10.1016/j.foodchem.2020.126248

45. Gitto L, Massini G, Mennini FS, Mento C, Buscema PM (2020) Affective symptoms and postural abnormalities as predictors of headache: an application of artificial neural networks. Neural Netw World 1:1–26

46. De Carlo M, Ferilli G, d'Angella F, Buscema M (2021) Artificial intelligence to design collaborative strategy: an application to urban destinations. J Bus Res 129:936–948

47. Buscema M, Ferilli G, Gustafsson C, Sacco PL (2022) Toward a precision, complexity-informed cultural policy design: structural bottlenecks to culture-led development in Skaraborg, Sweden. Commun Nonlinear Sci Num Simul 1007–5704. https://doi.org/10.1016/j.cnsns.2022.106886

48. Sisto R, Moleti A, Capone P, Sanjust F, Cerini L, Tranfo G, Massini G, Buscema S, Buscema PM, Chiarella P (2022) MicroRNA expression is associated with auditory dysfunction in workers exposed to ototoxic solvents and noise. Front Public Health Original Res. https://doi.org/10.3389/fpubh.2022.95818

49. Massimo B, Guido F, Christer G, Massini G, Luigi SP (2022) A nonlinear, data-driven, ANNs-based approach to culture-led development policies in rural areas: the case of Gjakove and Peć districts, Western Kosovo, Chaos, vol 162. Solitons & Fractals VI, p 112439. https://doi.org/10.1016/j.chaos.2022.112439

50. Buscema M, Torre FD, Massini G, Ferilli G, Sacco PL (2022) Reconstructing the emergent organization of information flows in international stock markets: a computational complex systems approach. Comput Econ. online 20 May 2022. https://doi.org/10.1007/s10614-022-10267-1

51. Erspamer C, Torre FD, Massini G, Ferilli G, Sacco PL, Buscema PM (2021) Global world (dis-) order? Analyzing the dynamic evolution of the micro-structure of multipolarism by means of an unsupervised neural network approach. Technol Forecast Soc Change. https://doi.org/10.1016/j.techfore.2021.121351. Received 22 March 2021; Received in revised form 16 September 2021; Accepted 8 November 2021

52. Buscema M, Ferilli G, Gustafsson C, Sacco PL (2023) Toward a precision, complexity-informed cultural policy design: structural bottlenecks to culture-led development in Skaraborg, Sweden. Commun Nonlinear Sci Num Simul 116:106886, 1007–5704. https://doi.org/10.1016/j.cnsns.2022.106886

53. Massimo B, Guido F, Christer G, Massini G, Luigi SP (2022) A nonlinear, data-driven, ANNs-based approach to culture-led development policies in rural areas: the case of Gjakove and Peć districts, Western Kosovo, Chaos. Solitons Fractals, VI, 162:112439. https://doi.org/10.1016/j.chaos.2022.112439

54. Bouchacourt D, Tomioka R, Nowozin S (2018) Multi-level variational autoencoder: learning disentangled representations from grouped observations. In: The thirty-second AAAI conference on artificial intelligence (AAAI-18), 2018, association for the advancement of artificial intelligence. www.aaai.org

55. Sarbu S, Volpi R, Peste A, Malago L (2018) Learning in Variational Autoencoders with Kullback-Leibler and Renyi Integral Bounds. arXiv:1807.01889v1 [cs.LG] 5 Jul 2018

56. Buscema M (1998) Recirculation neural networks. In: Buscema M (Ed), Substance use and misuse, vol 33, no 2. Special Issue on Artificial Neural Networks and Complex Social Systems, pp 383–388

57. Buscema M, Asadi-Zeydabady M, Lodwick W, Breda M (2015) The H0 function, a new index for detecting structural/topological complexity information in undirected graphs. Elsevier, Physica A 447(2016):355–378
58. Micciche' S, Bonanno G, Lillo F, Mantegna RN (2003) Degree stability of a minimum spanning tree of price return and volatility. Physica A: Statist Mech Appl 324(1–2):66–73. ISSN 0378-4371, https://doi.org/10.1016/S0378-4371(03)00002-5
59. Naidoo K, Massara E, Lahav O (2022) Cosmology and neutrino mass with the minimum spanning tree, vol 513, no 3. Monthly Notices of the Royal Astronomical Society, pp 3596–3609. https://doi.org/10.1093/mnras/stac1138. Published: 27 April 2022

Chapter 6
Attention, Consciousness, and Self-awareness

Based on the examination conducted, it appears that AI techniques are poised to demonstrate themselves as intelligent systems and possibly even sentient. In particular,

a. An extended statistical approximation of connections between words that can be generated in very long natural language texts (Large Language Model—LLM).
b. Superior pattern recognition capabilities compared to humans (sets of classifiers governed by Meta Net).
c. The ability of each node in an ANN to approximate holistic knowledge of all training data and thus develop focused attention based on the constraints of the given query (what we call *conscious nodes*).

However, an algorithmic ability of an artificial system to focus its attention based on context does not necessarily imply that the system can be considered conscious. It always involves probabilistic associations between what the system has learned and the themes of a new query it is subjected to, whether this calculation is performed through dot products, subsequently normalized ("Attention is all you need"), or via an ANN that operates spontaneously in this manner (Auto CM). Our points are different. Consciousness is hardly definable in a formal and rigorous manner. Furthermore, believing that the brain generates the mind assumes that the "mind" is an emergent effect of the behavior of a complex material system. In short, it is matter that, when it reaches certain levels of complexity, spontaneously generates what we experience as abstract, such as thoughts. If the brain were simple enough to understand, it would be too simple to be a brain.

This chapter presents a series of reasonable conjectures without formal proofs or experimental basis. It should be understood as a seminal contribution that may be a consequence of the preceding chapters. The responsibility for these ideas lies solely with one of the authors of this book (PMB) See Refs. [1–5].

© The Author(s), under exclusive license to Springer Nature Switzerland AG 2025   75
P. M. Buscema et al., *AI: A Broad and a Different Perspective*,
SpringerBriefs in Computational Intelligence,
https://doi.org/10.1007/978-3-031-80600-1_6

However, one can proceed as follows. An axiom is established, and it is verified whether the consequences of that axiom lead to verifiable events, lead to contradictions, and/or allow for the demonstration of useful theorems. In this case, the axiom could be the following. For a system to be conscious, it must be a living system.

Let's hypothesize that for a system to be living, it must be structurally imbalanced. For a system to be structurally imbalanced, it must have been designed in a world with D dimensions and then encapsulated in a world with D-1 dimensions. This way, the system will always feel "restless," that is, not at rest, as opposed to the condition of a cube resting on a plane. Therefore, this system will always seek to move, both physically and cognitively, to explore its minimum potential. However, since this system is more complex than the space into which it is immersed, it will never find a state of rest. For example, the heart beats because it is structurally imbalanced, the same happens with breathing, and the body undergoes a similar process (if it remains motionless for a long time, it develops bedsores). Living structures are thus forced to continuously adjust. That is, living structures are constantly in motion.

However, this structural restlessness is accompanied by a new fact. More complex living systems exhibit incredible mnemonic capacity, capacity for recall. For instance, an elderly human brain is capable of reactivating and reliving episodes from childhood. Now, it is known that after 50 years of life, a human subject has renewed all the atoms that made up their body in their early years. So, questions arise. "How does it remember events from its past if its material substrate at 50 years is entirely different from what it had at the age of ten?" "On which 'hard drive' or cloud of computers is its memory and information about its identity stored? The questions are not naive, as the neurons in our brain are also cells, and cells are ultimately made up of atoms. Therefore, when the "old" atoms give way to the "new" ones, how do the latter remember the functioning of the cells and connections that were constituted by the old atoms? Where is the information stored through which a subject continues to know "who they are," despite the passage of time and the global change in the matter of which every living subject is composed?

One possible answer that comes to mind is that the identity of every living subject is not based on "matter" and, ultimately, on its atoms at the moment, but rather on the trajectories and dynamic positions that these take within cells. New atoms continue to behave similarly to the old ones they replace in that place. But at this point, we must consider the hypothesis that the identity and memory of every living system are not only based on a material substances but on an immense set of abstract information (atom trajectories and relationships between atoms). These abstract (non-material but manifesting through matter) pieces of information should mark the specifics of identity and self-memory that characterize every complex living system.

One could argue that in living systems, assuming the initial axiom to be true and combining it with this new hypothesis, it is abstract information that generates matter-energy and not the other way around, as the mechanistic approach might suggest. The reason for this unusual inference is simple. A 4-dimensional structure, trapped in a 3-dimensional space, is always imbalanced and continuously adjusts. It moves, thinks, suffers, and so on because a part of the information in its structure cannot function in a space with one dimension less. When that excluded part of information presses to enter the 3D space, as a side effect, another part of its information is expelled. It is this feedback loop that operates as long as the system is alive. The information entering and leaving the system is always the same, but each time it enters from the 4 dimensions into the 3 dimensions from a different point, thus influencing the system to which it belongs differently. In other words, imagine the shadow of a cube crossing a 2D plane, each time at a different angle. It will draw a different shadow on the intersecting plane each time. Therefore, this missing part of information, in its attempt to enter 3D space, determines the present actions of the structure itself.

Every living structure, if these hypotheses are true, therefore, is characterized by a continuous transformation of information into matter-energy, where information represents the future trajectories that the 3-dimensional structure will follow. Let's provide an example in the form of a metaphor. Suppose there is a 2-dimensional world populated by 2-dimensional individuals. Now imagine a 3-dimensional bottle being flattened from the top to fit it into the 2-dimensional world at a constant speed. At time 1, the 2-dimensional individuals will see a shape emerging from nothing, and at time 2, they will see how the first shape is redrawn differently. They will then be inclined to believe that the new shape at time 2 is caused by the old shape that appeared at time 1: the past causes the present. The same reasoning can be iterated each time a part of the 3-dimensional bottle enters the 2D world. In reality, this reconstruction is incorrect because the actual dynamics are the opposite. It is the part of the bottle that has not yet entered the 2D world that causes the overall shape of the part that has already entered that world. In other words, in this example, it is the 3D "future" that creates the 2D present. We also emphasize that while the bottle moves in its original space at a constant speed, in the 2D world, the speed of movement of its parts experiences accelerations and decelerations that depend on the shape of the bottle itself, in fact, the derivatives of the bottle's shape at each point change.

One might also recall Plato's allegory of the cave in this context. The bodies of real 3D individuals living in the cave cast 2-dimensional shadows on its walls when at least one light source is present. Those who only observe the dynamics of those shadows might be inclined to believe that each shadow at time t_0—Delta is the complex cause of the shadow at time t_0. But this reasoning collapses and reveals its naivety when an additional dimension is added to the space in which this scene unfolds. It is the movements of real 3D individuals that cause the changes in the 2D shadows on the cave walls. Believing that the shadows cause the substance of the material forms of which they are projections is folly.

Ultimately, we are hypothesizing that just as in 3D space, matter projects shadows in 2D space, information in 4D space projects "shadows" in 3D space, which we recognize as matter and energy. After all, every living system possesses a specific genetic code, which, among other things, expresses itself over time in diversifying ways. This code is present, for example, in each of the approximately 4×10^{13} cells in a human body. However, the genetic information for each living structure is one. Following our hypothesis, genetic information in 4D space, being too complex to be immersed in 3D space, manifests in pieces as matter and energy in billions of partial 3-dimensional projections over time. Therefore, if the concept of "consciousness" implies that of a living system, then the homology, that is, having the same or a similar relation, associating a living system with a particular type of "machine" cannot function. But perhaps there's more. One of the most important effects of a living system is that it grows over time. At time 1, a cell divides into two cells, at time 2, there are four cells, at time 3, they become eight, and so on. In living systems, internal entropy decreases over time because the system becomes more complex and organized. In contrast, in "complicated" systems, which are often what we build, the passage of time results in spontaneous disorder, which will eventually, without external intervention, halt the system's functioning (an unmaintained plane eventually ceases to function).

What has been said implies that a living system in a 3-dimensional world (plus time) creates information (or displays new information) during its life cycle as its components change over time and reorganize over time. Therefore, time becomes, according to this hypothesis, the excess dimension specific to every living system, which manifests in the 3D world through information packages that attempt to complete the global information that characterizes that system.

In essence, every living system is a complex information structure in a 4-dimensional space that manifests itself in an ever-incomplete manner in 3-dimensional space as unstable matter, utilizing time to continuously adjust. As AI systems try to become more "human", they will need to come closer to these characteristics, if our hypotheses are correct.

References

1. D'Ariano GM, Faggin F (2022) Hard problem and free will: an information-theoretical approach. In: Scardigli F (eds) Artificial intelligence versus natural intelligence. Springer, Cham. https://doi.org/10.1007/978-3-030-85480-5_5
2. Faggin F (2021) Consciousness comes first. In: Kelly EF, Marshall P (eds) Consciousness unbound: liberating mind from the tyranny of materialism. Rowman & Littlefield, pp 283–322
3. Faggin F (2024) Irreducible: consciousness, life, computers, and human nature, paperback—31 May 2024
4. Perret D (2021) The intelligence behind 'nature'. Manifesting consciousness, nooks on demand GmbH. Paris, France
5. Kelly EF, Marshal P (eds) (2021) Consciousness unbound: liberating mind from the tyranny of materialism. Published by Rowman & Littlefield, London

Conclusion

The topics discussed in this essay may seem disconnected, yet a common thread rigorously connects them: in what direction is artificial intelligence useful for the development of humanity? We began by assuming that the overarching purpose of artificial intelligence is to construct systems capable of generating thoughts: from myriad cells made of known matter to chains of abstract concepts made of unknown "substances". This endeavor can serve at least two major purposes:

a. Constructing a new world where "gods" are only a few humans, those whom intelligent machines know how to construct. In this new world, other humans and new intelligent machines are the fluid, and thus interchangeable, elements that compose it and with which to play. This is a dangerous game because in a world where game elements are mere objects, free will becomes noise compared to any plan conceived by the game's creator. We are indeed talking about intelligent machines, which can make predictions and decisions surpassing the predictive abilities and decision-making speed of any human being. Reconstructing an artificial brain to better understand how ours functions helps us understand who we are and develop new possibilities for our lives and the environment. Using an artificial brain to replace ours helps us die sooner rather than defeat death.

b. Constructing tools that extend human capabilities to address the laws of nature still completely unviolated and incomprehensible by any science, those that allow for the spontaneous emergence of complex systems from the behavior of simpler ones: from atoms to molecules, from molecules to proteins, from proteins to cells, from cells to complex organisms; or if preferred: from electrochemical interactions between neurons to thought, from individual ants to their organized colonies, from individual animals and humans to communities, cities, institutions, from the simple transformation of hydrogen into helium to that universe whose distribution of stars, galaxies, and nebulae is still not fully explicable.

In this brief essay, we have delved into aspects of these two possible paths, highlighting the myths through which profit-oriented artificial intelligence has penetrated common opinion.

P. M. Buscema et al., *AI: A Broad and a Different Perspective*,
SpringerBriefs in Computational Intelligence,
https://doi.org/10.1007/978-3-031-80600-1

First, we have thus highlighted, through precise experiments, the fragility of the myth of "big data", and how human intelligence develops by eliminating vast amounts of useless data and selecting a few useful pieces of information. In this regard, we have also shown real-world applications of how important information can be hidden in few significant data (a single image or geographical coordinates of a process).

Subsequently, we have also demonstrated, through many experiments on real data, how many small and diverse Neural Networks can achieve much more accurate predictive results than a large and solitary neural network; we have also shown how excessive vertical depth of Convolutional Neural Networks risks collapsing the very principle of data-driven learning if not supported by horizontal complexity consisting of highly diverse adaptive algorithms, each expressing different opinions on the same data sample: the effectiveness of biodiversity produced by nature must be imitated in artificial intelligence through the "algodiversity" of machine learning systems. This is necessary if one wishes to understand how global behaviors emerge spontaneously through the interaction of their elementary components, which are inherently uninformative.

Finally, we have experimentally demonstrated how the dominant Neural Networks in the mainstream today follow purely Bayesian and/or frequentist calculations (who co-occurs with whom), while other types of Neural Networks, still little known, are capable of providing each of their elements with a holistic knowledge of the data on which they exhibit scale-invariant learning.

We have thus concluded our essay on the most relevant topic for a generalist artificial intelligence: can a system capable of exhibiting attentional abilities on the data it receives mark the initial step towards creating an artificial system conscious? Our response has been clear: systems with dynamic attentional mechanisms on incoming data represent an approximation of multiconditional probability calculation. Therefore, it is a more complex calculation than usual, but it is only a calculation that approximates quantities. Consciousness, we have hypothesized in the last paragraph of this essay, is a topological problem connected to the qualitative properties of a system, and not simply arithmetic, i.e., explainable through relations between quantities.

From these arguments, it emerges that an artificial intelligence aiming to understand how systems spontaneously increase their complexity with surprising qualitative leaps from one level to the next is the future of a science we currently lack. We have termed it the investigative approach to artificial intelligence. The reason for the name lies in its purpose: it attempts to understand the laws through which physical quantities generate abstract quantities we call information.

The simulation and emulation approaches we have discussed are undoubtedly useful and provide important knowledge and tools (algorithms) even for the development of the investigative approach: there is nothing that does not deserve to be unveiled. However, the principle of not destroying what one is unable to rebuild always holds true.

Therefore, when the primary objective of a certain artificial intelligence, based on brute force and economic profit, becomes to replace nature and the human brain in fields where human responsibility, free will, and natural processes cannot be delegated to automatic decisions, then transforming into "gods" of a world that is still not fully understood is an unwitting plan for self-destruction: from manure, flowers continue to bloom; from circuits, nothing emerges.